NOT IN CODE, POSSIBLY ADOPTED. WOULD NOT ⌐ ∪
  GD.
  TOXIC GASES
" LEVEL OF CONCERN " = 0.1 ⟩
MATERIAL HAZARD INDEX = M

  CLASS I REGULATED MATER   ⌐,000
  CLASS II   "   "   MHI   10,000 ~ 500,000
  CLASS III   "   "   4,900 ~ 10,000

EXEMPT AMOUNT   POISON A = 1/4 lb.
                 OTHER REG = 1 lb.

CATASTROPHIC RELEASE : dilution   1/2 IDLH @ STACK
EXHAUSTED.

LEVELS OF CONCERN

 — CORPORATE POLICY
 — INDUSTRY STANDARDS BEST AVAILABLE PRACTICE

 — EEOC REGULATIONS
 — ADA - civil code
 — MODEL CODES — LEGAL REQUIREMENT

CODE COMPLIANCE FOR ADVANCED TECHNOLOGY FACILITIES

CODE COMPLIANCE FOR ADVANCED TECHNOLOGY FACILITIES

A Comprehensive Guide for Semiconductor and Other Hazardous Occupancies

by

William R. Acorn

Acorn Engineering & Consulting
Tucson, Arizona

np **NOYES PUBLICATIONS**
Park Ridge, New Jersey, U.S.A.

Library of Congress Catalog Card Number: 93-28549
ISBN: 0-8155-1338-0
Printed in the United States

Published in the United States of America by
Noyes Publications
Mill Road, Park Ridge, New Jersey 07656

10 9 8 7 6 5 4 3 2 1

Library of Congress Cataloging-in-Publication Data

Acorn, William R.
    Code compliance for advanced technology facilities : a
comprehensive guide for semiconductor and other hazardous
occupancies / by William R. Acorn.
        p.   cm.
    Includes index.
    ISBN 0-8155-1338-0
    1. Hazardous substances--Safety measures--Handbooks, manuals, etc.
2. Hazardous substances--Law and legislation--United States--
Handbooks, manuals, etc.   I. Title
    T55.3.H3A27   1993
    363.17'6'0973--dc20   93-28549
                                                         CIP

MATERIALS SCIENCE AND PROCESS TECHNOLOGY SERIES

Editors

Rointan F. Bunshah, University of California, Los Angeles *(Series Editor)*
Gary E. McGuire, Microelectronics Center of North Carolina *(Series Editor)*
Stephen M. Rossnagel, IBM Thomas J. Watson Research Center
(Consulting Editor)

Electronic Materials and Process Technology

HANDBOOK OF DEPOSITION TECHNOLOGIES FOR FILMS AND COATINGS, Second Edition: edited by Rointan F. Bunshah

CHEMICAL VAPOR DEPOSITION FOR MICROELECTRONICS: by Arthur Sherman

SEMICONDUCTOR MATERIALS AND PROCESS TECHNOLOGY HANDBOOK: edited by Gary E. McGuire

HYBRID MICROCIRCUIT TECHNOLOGY HANDBOOK: by James J. Licari and Leonard R. Enlow

HANDBOOK OF THIN FILM DEPOSITION PROCESSES AND TECHNIQUES: edited by Klaus K. Schuegraf

IONIZED-CLUSTER BEAM DEPOSITION AND EPITAXY: by Toshinori Takagi

DIFFUSION PHENOMENA IN THIN FILMS AND MICROELECTRONIC MATERIALS: edited by Devendra Gupta and Paul S. Ho

HANDBOOK OF CONTAMINATION CONTROL IN MICROELECTRONICS: edited by Donald L. Tolliver

HANDBOOK OF ION BEAM PROCESSING TECHNOLOGY: edited by Jerome J. Cuomo, Stephen M. Rossnagel, and Harold R. Kaufman

CHARACTERIZATION OF SEMICONDUCTOR MATERIALS, Volume 1: edited by Gary E. McGuire

HANDBOOK OF PLASMA PROCESSING TECHNOLOGY: edited by Stephen M. Rossnagel, Jerome J. Cuomo, and William D. Westwood

HANDBOOK OF SEMICONDUCTOR SILICON TECHNOLOGY: edited by William C. O'Mara, Robert B. Herring, and Lee P. Hunt

HANDBOOK OF POLYMER COATINGS FOR ELECTRONICS, 2nd Edition: by James Licari and Laura A. Hughes

HANDBOOK OF SPUTTER DEPOSITION TECHNOLOGY: by Kiyotaka Wasa and Shigeru Hayakawa

HANDBOOK OF VLSI MICROLITHOGRAPHY: edited by William B. Glendinning and John N. Helbert

CHEMISTRY OF SUPERCONDUCTOR MATERIALS: edited by Terrell A. Vanderah

CHEMICAL VAPOR DEPOSITION OF TUNGSTEN AND TUNGSTEN SILICIDES: by John E. J. Schmitz

ELECTROCHEMISTRY OF SEMICONDUCTORS AND ELECTRONICS: edited by John McHardy and Frank Ludwig

v

HANDBOOK OF CHEMICAL VAPOR DEPOSITION: by Hugh O. Pierson

DIAMOND FILMS AND COATINGS: edited by Robert F. Davis

ELECTRODEPOSITION: by Jack W. Dini

HANDBOOK OF SEMICONDUCTOR WAFER CLEANING TECHNOLOGY: edited by Werner Kern

CONTACTS TO SEMICONDUCTORS: edited by Leonard J. Brillson

HANDBOOK OF MULTILEVEL METALLIZATION FOR INTEGRATED CIRCUITS: edited by Syd R. Wilson, Clarence J. Tracy, and John L. Freeman, Jr.

HANDBOOK OF CARBON, GRAPHITE, DIAMONDS AND FULLERENES: by Hugh O. Pierson

Ceramic and Other Materials—Processing and Technology

SOL-GEL TECHNOLOGY FOR THIN FILMS, FIBERS, PREFORMS, ELECTRONICS AND SPECIALTY SHAPES: edited by Lisa C. Klein

FIBER REINFORCED CERAMIC COMPOSITES: edited by K. S. Mazdiyasni

ADVANCED CERAMIC PROCESSING AND TECHNOLOGY, Volume 1: edited by Jon G. P. Binner

FRICTION AND WEAR TRANSITIONS OF MATERIALS: by Peter J. Blau

SHOCK WAVES FOR INDUSTRIAL APPLICATIONS: edited by Lawrence E. Murr

SPECIAL MELTING AND PROCESSING TECHNOLOGIES: edited by G. K. Bhat

CORROSION OF GLASS, CERAMICS AND CERAMIC SUPERCONDUCTORS: edited by David E. Clark and Bruce K. Zoitos

HANDBOOK OF INDUSTRIAL REFRACTORIES TECHNOLOGY: by Stephen C. Carniglia and Gordon L. Barna

CERAMIC FILMS AND COATINGS: edited by John B. Wachtman and Richard A. Haber

Related Titles

ADHESIVES TECHNOLOGY HANDBOOK: by Arthur H. Landrock

HANDBOOK OF THERMOSET PLASTICS: edited by Sidney H. Goodman

SURFACE PREPARATION TECHNIQUES FOR ADHESIVE BONDING: by Raymond F. Wegman

FORMULATING PLASTICS AND ELASTOMERS BY COMPUTER: by Ralph D. Hermansen

HANDBOOK OF ADHESIVE BONDED STRUCTURAL REPAIR: by Raymond F. Wegman and Thomas R. Tullos

CARBON–CARBON MATERIALS AND COMPOSITES: edited by John D. Buckley and Dan D. Edie

CODE COMPLIANCE FOR ADVANCED TECHNOLOGY FACILITIES: by William R. Acorn

Preface

Industries which utilize hazardous materials and cleanrooms are seldom static; they are either expanding, rearranging or changing manufacturing processes. Facilities which utilize hazardous liquids and gases represent a significant potential liability to the owner, operator and general public in terms of personnel safety and preservation of assets.

This book was conceived to give the reader a guide to understanding the requirements of the various codes and regulations that apply to the design, construction and operation of facilities utilizing hazardous materials in their processes. Spawned many years ago as a guide to index and cross-reference the codes related to semiconductor wafer fabrication facilities, it was intended as a means for the author's design teams to readily assess the issues and develop compliance strategies. It has evolved over the years to provide examples from real world experiences and practical application of regulations to clients' facilities.

To some extent, this handbook continues to target the very specific and unique requirements of semiconductor wafer fabrication facilities; however, it is applicable to a wide variety of other hazardous occupancies. It is our aim to provide you with a useful reference to simplify and expedite your code compliance research no matter what type of hazardous processes / facilities you are involved with.

This book will give you an awareness of the requirements and potential ramifications of compliance with regulations pertaining to advanced technology facilities. No general discussion of the "H" occupancy codes can give you all of the answers for every situation. You must review the issues and develop a unique compliance strategy for each facility. In addition, the unique nature of facilities using hazardous materials will virtually ensure the need for interpretive application of the regulations to any given situation. Thus, the local code jurisdiction will help to shape the compliance strategy to your specific project.

Excerpts from various codes are provided in the book for the reader's convenience. They are <u>not</u>, nor are implied to be, a substitute for your research into the full body of all the code documents. Where portions of the code text have been omitted for the purpose of brevity, we have attempted to indicate the discontinuity with '...' notations. Where text is underlined or highlighted, this is generally done to emphasize <u>our perception</u> of key words or phrases. Such emphasis should not be construed to limit the full meaning or content of the code requirements. Where there is room for interpretation, consultation with the local code official is encouraged—after you have developed a compliance concept for their evaluation.

Cross references have been provided for the reader's convenience to help locate citations in various code sections pertaining to issues of a similar nature. The cross references have been carefully considered; however, they may not be all-inclusive in scope, and the interpretation of local jurisdictions will vary such that other relationships of requirements are dictated. The reader should consider the lists of applicable codes as a guideline and starting point in his or her analysis of a given situation.

This book generally follows the requirements of the 1991 editions of the "model" documents of the *Uniform Building Code, Uniform Fire Code* and *National Fire Protection Association*. Other codes, such as the *BOCA (Building Officials and Code Administrators) National Building Code* and the *Standard Building Code* (of the Southern Building Code Congress) may apply in your area. Where they do, you must consult them for specific provisions which vary from those of the UBC and UFC "family". In any case, local amendments of regulators in your area may vary markedly from the national "model codes." Again, **the reader is responsible for using the book with extreme caution where a specific application, design or operation is concerned.** Communication and liaison with the local code regulators is mandatory to your success!

The nature of codes related to hazardous occupancies is changing. As more information becomes available to operators and regulators, new requirements are promulgated. Future editions of this handbook will reflect changes in regulations and in technology, as they occur.

In the event that you disagree with our interpretation of an issue or find an error, we will welcome your input! Likewise, if you have any suggestions for additional topics or other improvements for future editions, we encourage you to contact the author or publisher.

We appreciate your use of this book and sincerely hope it will benefit you in developing code compliance strategies for the very complex and challenging advanced technology hazardous facilities. We wish you the best of success in your endeavors!

Tucson, Arizona   William R. Acorn, P. E.
August, 1993   Principal, Acorn Engineering & Consulting

Acknowledgements

In this process of writing my first book, I have learned many things: the virtue of persistence, the value of communication, and the humility of learning how many things I really didn't fully understand. I realize the benefit and joy of having friends, associates, clients and colleagues upon whom I could rely for information and encouragement opportunities. No endeavor of this magnitude is the fruit of only one person; it is made possible by many. In my own way, I would like to take this opportunity to thank some of those people who have helped make this book a reality.

As the process of writing this book unfolded, I soon realized that I was not an "expert" in the subject, merely a design practitioner with a desire to pull together, as a comprehensive whole, the many bits and pieces of information I had obtained over time. There have certainly been many before me who have contributed much to the body of knowledge surrounding the design and construction of hazardous occupancies. I am grateful to Larry Fluer and Reinhard Hanselka for their pioneering and on-going work in this field. I am also grateful to the many dedicated code professionals I have had the pleasure of working with. While it wouldn't be possible to name them all, some who come to mind include Doug Hood, Tom Hedges, Jim Harl, Mike Benoit, Clarence Hainey, Scott Stookey, Jim Singleton and Fred Blackmore.

The ability to practice consulting engineering and develop expertise in the design and analysis of hazardous facilities depends on clients who are confident in your ability and retain you to provide services. I have been extremely blessed in my career to have served some of the best in the industry. Although I cannot list them all, some have been particularly supportive and instrumental to my success. A leader in the development of facilities and products, Motorola is our most significant client with facilities engineering leaders Bob Predmore, Fred Nea, Arnie Graham, John Snyder, Mike Scott, Dean Hooks, Rich Weigand, Roger Woods and Bob Jones, to name a few. Intel is another industry leader I have been privileged to serve, with the support of Art Stout, Brian Woods, Dennis Emerson and Gil Hansen. Other important clients and friends include Dan Dierken, Tim Smith, Leo Pachek and Lee Carlstrom of Northern Telecom, Duane Kiihne of IBM, Warren Johnson of 3M, Nina Vulpetti of Ford Motor Company, Steve Burnett and Tommy Thompson of SEMATECH.

Many vendors and sales representatives have assisted myself and my firm by sharing their knowledge of the state-of-the-art and developments in their respective industries. A few have been particularly instrumental in the progress of this book, including Dave Jones and Chris Carlson of Air Products, Paulette Spaunburg and Jeff Conrad of Semi-Gas, John Traub of Systems Chemistry, Ernie Power III of Power-Com, Dave Sorenson of Empire Power Systems/Caterpillar, Fred Serkasevich of Simplex, Terry White of Telosense, etc., etc., etc.

Ted Bielli of Advanced Technology Associates deserves special recognition, because it was Ted who first encouraged me to have what was then a loose-leaf notebook enhanced and published into a "real" reference book. Ted introduced me to the publisher, George Narita of Noyes Publications.

This book would not have been completed without the assistance and input of my colleagues at Acorn Engineering and Consulting. Specifically, Stephen Chansley, John Hall, Bob Muzzy, Bob Alcalá, Lee White and Eugene Moreno. Janice Marshall patiently word-processed the revisions and the revisions to revisions, adding ideas on improving the readability. Al Gastellum assisted with the development of some of the graphics. Last and certainly not least, Gina Gentry McElroy was the catalyst responsible for ensuring that I stick with this endeavor, even at times when my enthusiasm was much less than high and I lacked the energy to carry it to completion. Gina's gentle, yet persistent, encouragement helped get me to this point and I am very thankful.

My wife, and children have inspired me to be the best I can be—professionally and as a person. I thank them for their patience and support when I was working to keep an active consulting engineering practice going, when I should have been spending time with them.

August, 1993   William R. Acorn

NOTICE

The writers and editors of this book cannot assume any liability with regard to the misapplication or misinterpretation of any code provision which may result from the use of the book. The user is solely responsible for application of the principles and requirements of the codes and other regulations to each specific set of circumstances. You are encouraged to discuss your proposed compliance strategies with the local Authorities Having Jurisdiction over your project, early on in the process

To the best of our knowledge the information in this publication is accurate; however the Publisher does not assume any responsibility or liability for the accuracy or completeness of, or consequences arising from, such information. This book is intended for informational purposes only. Mention of trade names or commercial products does not constitute endorsement or recommendation for use by the Publisher. Final determination of the suitability of any information or product for use contemplated by any user, and the manner of that use, is the sole responsibility of the user. We recommend that anyone intending to rely on any recommendation of materials or procedures mentioned in this publication should satisfy himself as to such suitability, and that he can meet all applicable safety and health standards.

Table of Contents

1

Introduction to Codes

1.1   HISTORY OF THE HAZARDOUS OCCUPANCY CODES

Codes are ordinances (laws) adopted by the local government jurisdiction to establish minimum guidelines for construction and/or operation of a facility to protect life and property.   The codes are legal documents, which are enforceable to the extent that the building official has the right (with reasonable cause) to enter the premises, inspect, declare a facility unsafe and close it.   The earliest known building code is that of Hammurabi, founder of the Babylonian Empire (*Refer to* **Figure 1.1**).

Since 1980, a dramatic transition has taken place in the governing codes for facilities utilizing hazardous materials.   Prior to 1985, industrial occupancies utilizing hazardous materials were designed under the provisions of the Uniform Building Code (UBC) Group B-Division 2.

In 1985, the Uniform Building Code and the Uniform Fire Code (UFC) first published Section 911 and Article 51 respectively.   These codes specifically address building and operating requirements for semiconductor fabrication facilities.   The provisions for facilities related to the semiconductor industry were more clearly (and prescriptively) outlined in the 1988 edition of the model codes.   In 1991, as the typical three-year model code cycle continued, the UBC/UFC and other codes were again formally modified.   This book includes citations and revisions based upon the 1991 editions.

1

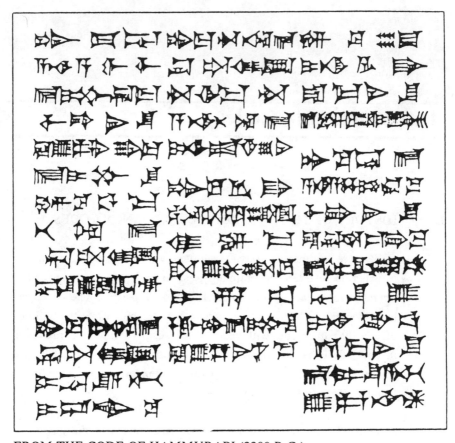

FROM THE CODE OF HAMMURABI (2200 B.C.)

If a builder builds a house for a man and does not make its construction firm and the house collapses and causes the death of the Owner of the house -- that builder shall be put to death. If it causes the death of a son of the Owner -- they shall put to death a son of that builder. If it causes the death of a slave of the Owner -- he shall give to the Owner a slave of equal value. If it destroys property -- he shall restore whatever it destroyed and because he did not make the house firm, he shall rebuild the house which collapsed at his own expense. If a builder builds a house and does not make its construction meet the requirements and a wall falls in -- that builder shall strengthen the wall at his own expense.

Figure 1.1   THE CODE OF HAMMURABI, founder of the Babylonian Empire, is the earliest known code related to buildings.

As shown in **Figure 1.2** (*see next page*), the dangers involved in the design and operation of a facility utilizing hazardous materials are very real. The potential loss of life and property cannot be taken lightly. This book is dedicated to enhancing your understanding of the issues and helping to ensure that you do not get in the position of looking at your facility in this condition.

1.2   OVERVIEW OF APPLICABLE CODES

Today, the basic "model" codes which affect hazardous facilities are numerous. In the United States, the following generally apply:

- Uniform Building Code (UBC)
- Uniform Mechanical Code (UMC)
- Uniform Plumbing Code (UPC)
- Uniform Fire Code (UFC)
- National Fire Protection Association (NFPA)
- National Electrical Code (NEC) (NFPA 70)
- Life Safety Code (NFPA 101)
- Protection of Cleanrooms (NFPA 318)
- Applicable Local Codes
- Local Pollution Abatement and Environmental Ordinances
- Underwriters Laboratories (U.L.)
- Federal Ordinances (such as OSHA and EPA regulations)
- Building Officials and Code Administrators International's National Building Code (BOCA)
- BOCA's National Fire Prevention Code (NFPC)
- Southern Building Code Congress International's Standard Building Code (SBCC)

The latest versions of the UBC and UFC codes are the 1991 editions, and most jurisdictions are either in the process of adopting, or have adopted, these editions. Your jurisdiction may still be operating under the 1988 editions; therefore, as you begin a new project, you should discuss the correct code edition to use as a basis for evaluation. In most cases, the 1991 edition will be either accepted, or preferred, even if not formally adopted.

Figure 1.2 THE HAZARDS ARE REAL; Recent Fire at Printed Circuit Board Facility, Phoenix, Arizona. (AP Photo)

In the eastern and southern states of the U.S., the Uniform Building and Fire Codes generally are not recognized. In the midwest and east, the BOCA (Building Officials and Code Administrators International, Inc.) National Building Code generally applies. In the southeastern portion of the U.S., the SBCC (Southern Building Code Congress International) Standard Building Code generally applies.

Prior to the 1993 edition, the BOCA National Code was virtually silent with respect to provisions governing the storage and use of hazardous materials. In the 1993 edition, definitions, requirements and provisions concerning hazardous occupancies have been added that closely parallel those of the UBC and UFC. The SBCC has carried provisions concerning hazardous materials since the 1991 edition.

Due to the fact that the UBC/UFC regulations have been in existence longer and are more extensive than those of BOCA or SBCC concerning hazardous occupancies, this book focuses on the UBC/UFC family of codes. If your project or facility is located in an area governed by one of the other codes, or is governed by codes which do not address hazardous occupancies, it will be wise to consider using the UBC/UFC requirements and the discussions in this book as guidelines to design, construction and operation. In the absence of law governing your facility, it may be your legal responsibility and it certainly is your moral responsibility, to be aware of, consider and implement the best available practices which reasonably apply to the facility and its operation.

The codes create an intricate web of requirements for the design and operation of industrial facilities. In order to gain perspective on the relationship of the various codes to one another and their overall application to the facility, it will be helpful to examine the stated scope and/or purpose of some of the regulations:

UBC 102 Purpose

"The purpose of this code is to <u>provide minimum standards to safeguard life or limb, health, property and public welfare</u> by regulating and controlling the design, construction, quality of materials, use and occupancy, location and maintenance of all buildings and structures within this jurisdiction and certain equipment specifically regulated herein..."

UFC 51.101 Semiconductor Fabrication Facilities using HPMs - Scope

"<u>Group H, Division 6</u> Occupancies shall be in accordance with this article. The storage, handling and use of hazardous production materials (HPM) shall be in accordance with this article and other applicable provisions of this code. Required devices and systems shall be <u>maintained</u> in operable condition."

UFC 79.101(a) Flammable and Combustible Liquids - Scope

"Storage, use, dispensing, mixing and handling of flammable and combustible liquids shall be in accordance with this article, except as otherwise provided in other laws or regulations... When flammable or combustible liquids present multiple hazards, all hazards shall be addressed. See Article 80..."

UFC 80.101(a) Hazardous Materials - Scope

"Prevention, control and mitigation of dangerous conditions related to storage, dispensing, use, and handling of hazardous materials and information needed by emergency response personnel shall be in accordance with this article..."

(Note: Remember that these are <u>excerpts</u> from the codes. Do not rely solely on the citations provided here, but always refer to the full code text on any issue!)

In general, the UBC governs <u>design of facilities</u>, while the UFC governs <u>use and/or operation and maintenance of facilities, as well as design</u>.   The line between the two documents and other applicable codes is not clearly drawn on all issues and, in some cases, several codes address the same kinds of requirements.   It is the stated objective of both the UBC and UFC that the documents be complimentary and without conflict.   To this end, the 1991 editions of each have reduced the amount of overlap in their text.   In the 1991 editions of these codes, you will find references to the other code, rather than repeating the requirements.

In addition, the following must also be considered:

- Corporate Safety Directives
- Insurance Underwriter Requirements

Codes state the <u>minimum design requirements</u>; therefore, they may not limit your legal or moral obligations.   Additional design features or construction which enhance the safety and minimize the hazard to personnel, property and the public must be considered when they can be reasonably implemented.   Again, the principle of "best available practice" should be considered in each situation, particularly in industries or facilities in which the technology (and potential hazards) are changing faster than the codes can keep up.

<u>Other regulations</u> apply to the gamut of hazardous and nonhazardous buildings.   Some of these which are not addressed in this book include:

ADA   -   Americans With Disabilities Act
OSHA -   Occupational Safety and Health Administration regulations

1.3   OVERVIEW OF UNIFORM BUILDING CODE

The Uniform Building Code and related codes, such as the Uniform Fire Code and the National Fire Protection Association Pamphlets, are "model" codes developed by various agencies in the United States. Each code jurisdiction, whether it be a city, county or state, has the opportunity to adopt these model codes in whole or in part. Generally, the local jurisdiction makes amendments which revise provisions of various code sections to reflect local requirements.

The Uniform Building Code is published by the International Conference of Building Officials (ICBO). The UBC is the most widely accepted and adopted of the "model" codes in the country.

The Uniform Building Code is frequently referred to as an individual document, however, the ICBO codes also include the Uniform Mechanical Code (UMC) and the Uniform Plumbing Code (UPC). The UBC family of documents will be referred to as simply the UBC, however, realize that we are also talking about the related mechanical and plumbing codes. The National Electrical Code (NEC) originates in the National Fire Protection Association (NFPA) family of codes as Pamphlet 70 of the NFPA series.

The UBC is divided into the following parts:

- Part I covers administrative provisions and documents the origin, organization and enforcement procedures of the code.

- Part II covers definitions and abbreviations used throughout the code. While this part may be considered by many to be non-essential, it is important that the reader of the code understand the specific definition of any term used in the code for proper interpretation of the code intent. As an example, Chapter 4 defines "Service Corridor," "Emergency Control Station," "Hazardous Production Material," etc. as follows:

"SERVICE CORRIDOR is a fully-enclosed passage used for transporting hazardous production materials and for purposes other than required exiting."

"EMERGENCY CONTROL STATION is an approved location on the premises of a Group H, Division 6 Occupancy where signals from emergency equipment are received and which is continually staffed by trained personnel."

"HAZARDOUS PRODUCTION MATERIAL (HPM) is a solid, liquid or gas that has a degree of hazard rating in health, flammability or reactivity of 3 or 4 and which is used directly in research, laboratory or production processes which have, as their end product, materials which are not hazardous."

Again, proper interpretation of the code goes beyond a general understanding of the words used in the code and requires the reader of the code to understand the context in which any specific word is used.

- Part III of the code includes Chapters 5 through 16 and defines occupancy requirements. Of specific interest to this discussion are Chapter 5 - General Requirements for All Occupancies, Chapter 7 - Requirements for Group B Occupancies and Chapter 9 - Requirements for Group H Occupancies. Group H Occupancies are generally considered hazardous. Within each occupancy group are a number of classes which further define the use of the facility and the specific requirements of the code for construction and operation.

- Part IV of the code relates to requirements based on the various types of construction and construction materials used in buildings. The types of construction represent varying degrees of public safety and resistance to fire. There are five general categories or types of building construction defined in the code in Table 17-A.

■ Part V of the code includes Chapters 23 through 28 and defines, in detail, structural engineering requirements of various types of buildings construction. This is the most technically-oriented part of the code and is extremely prescriptive in nature.

■ Part VI outlines detailed regulations for general construction requirements. Such issues as excavation, veneer, roof construction, exiting, and fire-extinguishing systems are covered. Of particular interest is Chapter 33, which covers detailed exiting requirements from the various types of buildings and various occupancies. Chapter 38 includes provisions for fire-extinguishing systems within buildings related to automatic fire suppression and sprinkling systems. A new Chapter 31 deals with the requirements for accessibility of the handicapped or disabled.

■ Part VII includes Chapters 42 and 43 on fire-resistive and fire-protection standards of building construction.

■ Part VIII includes regulations for the use of public streets and public property, specifically during construction and concerning permanent occupancy of public property.

■ Part IX and Part X cover specific subjects such as wall coverings, prefabricated construction, elevators, skylights, glass, glazing, etc.

■ Part XI is entitled UNIFORM BUILDING CODE STANDARDS and includes, by reference, the UBC standards and details which are compiled in a separate volume. The UBC standards are a good reference and valuable source document for any facility owner or designer.

Without diminishing the need to consider all requirements of the code, the primary chapters in the UBC which are of importance in the design and construction of a hazardous manufacturing occupancy include:

Chapter 5	Occupancy Classification and Requirements
Chapter 9	Hazardous Occupancies (Group H)
Chapter 17	Types of Building Construction Classifications and Requirements
Chapter 33	Exit Requirements
Chapter 43	Fire-Resistive Construction Requirements

The UBC excerpt following, entitled "Effective Use of The Uniform Building Code," may assist the reader in the general understanding of the relationship of the various sections of the UBC.

EFFECTIVE USE OF THE UNIFORM BUILDING CODE
(Excerpts from the UBC)

"The following procedures may be helpful in using the Uniform Building Code:

1.   Classify the building:

A.   OCCUPANCY CLASSIFICATION: ...Determine the occupancy group which the use of the building most nearly resembles.   See the 01 sections of Chapters 5 through 12. See Section 503 for buildings with mixed occupancies.

B.   TYPE OF CONSTRUCTION: Determine the type of construction of the building by the building materials used and the fire resistance of the parts of the building.   See Chapters 17 through 22.

C.   LOCATION ON PROPERTY: Determine the location of the building on the site and clearances to property lines and other buildings from the plot plan.   See Table 5-A and 03 sections of Chapters 18 through 22 for exterior wall and wall opening requirements based on proximity to property lines. See Section 504 for buildings located on the same site.

D.   ALLOWABLE FLOOR AREA: Determine the allowable floor area of the building.   See Table No. 5-C for basic allowable floor area based on occupancy group and type of construction.   See Section 506 for allowable increases based on location on property and installation of an approved automatic fire-sprinkler system.   See Section 505 (b) for allowable floor area of multi-story buildings.

E.   HEIGHT AND NUMBER OF STORIES: Compute the height of the building, Section 409 and determine the number of stories, Section 420.   See Table 5-D for the maximum height and number of stories permitted based on occupancy group and type of construction.   See Section 507 for allowable story increase based on the installation of an approved automatic fire-sprinkler system.

2.   Review the building for conformity with the occupancy requirements, in Chapters 6 through 12.

3.   Review the building for conformity with the type of construction requirements, in Chapters 17 through 22.

4.   Review the building for conformity with exiting requirements in Chapter 33.   (See Chapter 31 where access for the disabled is required.)

5.   Review the building for other detailed code regulations in Chapters 29 through 54, Chapter 56 and the Appendix.

6.   Review building for conformity with engineering regulations and requirements for materials of construction.   See Chapters 23 through 28."

1.4   OVERVIEW OF UNIFORM FIRE CODE

1.4.1   General: The Uniform Fire Code is published by the ICBO and Western Fire Chiefs Association. In general, buildings are designed and constructed under the provisions of the <u>Uniform Building Code</u>, whereas, the <u>Uniform Fire Code</u> covers the maintenance and operation of facilities, as well as design and construction requirements. The 1991 edition of the UFC, as with the UBC, has been modified to improve the correlation of the two documents, with the intent that there be no conflict between them. Neither the UBC nor UFC are intended to stand alone; rather, they must be used as a comprehensive program of codes.

1.4.2   Organization of the UFC: The UFC is divided into eight (8) basic parts as follows. Articles relevant to the design, construction and use of hazardous occupancies are highlighted and discussed.

PART I   Administrative covers the scope, organization and authority of the regulation.

PART II   Definitions and Abbreviations

PART III   General Provisions for Fire Safety covers fire suppression facilities, maintenance of egress and accessibility and emergency procedures.

PART IV   Special Occupancy Uses covers the requirements for specific types of facilities.

PART V   Special Processes

                          Article 51 - Semiconductor Fabrication Facilities Using Hazardous Production Materials

PART VI   Special Equipment

PART VII   Special Subjects covers the requirements for use and maintenance of specific materials including:

                          Article 74 - Compressed Gases

                          Article 75 - Cryogenic Fluids

                          Article 79 - Flammable and Combustible Liquids

                          Article 80 - Hazardous Materials

PART VIII   Appendices
                    Division I - Existing Buildings
                    Division II - Special Hazards
                                  II-A.   Suppression and Control
                                             of Hazardous Fire Areas
                                  II-E.   Hazardous Material
                                             Management Plan
                                             and Hazardous Material
                                             Inventory Statements
                    Division III - Fire Protection
                    Division VI - Informational
                                  VI-A.   Hazardous Materials
                                             Classifications
                                  VI-D.   Reference Tables from the
                                             Uniform Building Code

The articles of greatest interest to our discussion of hazardous occupancies are:

Article 51   Semiconductor Facilities
Article 79   Flammable and Combustible Liquids
Article 80   Hazardous Materials

1.4.3   **Article 51:** Article 51, "Semiconductor Fabrication Facilities Using Hazardous Production Materials," was first published in the Uniform Fire Code in the 1985 edition. This section of the code was the first to be developed utilizing a risk management approach.

By 1981, both the International Conference of Building Officials and the Western Fire Chiefs Association had become aware of the need to develop a code which would regulate the semiconductor industry. The industry had developed over the previous decade under building and fire codes which did not recognize the technological advances of the industry or its unique needs. In an effort to address those problems, a sub-committee of the ICBO Fire and Life Safety Committee was convened.

This subcommittee consisted of representatives of the Uniform Building Code, the Uniform Fire Code and semiconductor manufacturers. The group met regularly in San Jose, California throughout 1982 and 1983 and presented its recommendations to the Western Fire Chiefs Association at its annual conference in 1984.

Article 51 corresponds most directly to Chapter 9 of the Uniform Building Code, governing hazardous occupancies and, more specifically, to Section 911. These two documents will be used hand-in-hand throughout the course of this book. Perhaps the most significant connection between the two documents is in the method of dealing with the permissible quantity of HPMs in an "H" occupancy. The UBC governs the "exempt" amount of HPM, or that quantity below which the facility is not considered an "H" occupancy. Article 51 governs the <u>maximum</u> quantity of HPMs which may be contained in a single H-6 occupancy.

1.4.4 **Article 80:** As is the case for Article 51, Article 80 "Hazardous Materials" was first introduced in the 1985 edition. In 1985, Article 80 was five pages long. Article 80 is the most prescriptive of all the codes related to facilities utilizing hazardous materials. Design and operation of such a facility must carefully respond to its requirements. The text of Article 80 has grown to almost 50 pages of highly technical material.

Article 80 applies to all hazardous materials including those regulated elsewhere in the codes. When specific requirements concerning a material are provided in other articles, those specific requirements may supersede Article 80. (As an example, Article 79 "Flammable and Combustible Liquids" may have more specific discussion about these types of materials than Article 80.)

The intent is to <u>prevent, control and mitigate dangerous conditions</u> related to hazardous materials. In addition to containing administrative and engineering controls, it is intended to provide information needed by emergency responders. Article 80 is intended to be used in conjunction with the UBC and is a companion document to Chapter 9 of the UBC. As with all codes, Article 80 provides minimum standards and should not be construed to limit or negate the need for more stringent controls in specific applications.

The provisions of Article 80 related to health hazards may be waived when the chief or other official charged with the enforcement of the code has determined that such enforcement is preempted by other codes, statutes or ordinances (UFC 80.101(c)).

UFC Article 80 is organized into the following divisions:

Division I - General Provisions.   This division provides general requirements for the storage, dispensing, use and handling of hazardous materials and gives definitions to be used throughout the article. This article also directs that, where required by "the chief", a Hazardous Materials Management Plan and a Hazardous Inventory Statement be prepared in accordance with Appendix II-E of the UFC. Most jurisdictions currently require the filing and maintenance of both a management plan and an inventory statement. The article also governs the release and unauthorized discharge of hazardous materials.

Division II - Classification by Hazard.   This division divides hazardous materials into two hazard categories which are directly correlated to Tables 9-A and 9-B of the UBC:

    a)   Physical Hazards
    b)   Health Hazards

Physical hazards primarily present a hazard of fire or explosion, but may also present a health hazard. Health hazards primarily present a short- or long-term exposure problem to humans, animals, or the environment, but may also present a physical hazard. When hazardous materials present multiple hazards, all hazards must be addressed in accordance with the provisions of this Article.

Division III - Storage Requirements. This division applies to the storage of hazardous materials in excess of exempt amounts (specified in each section). General provisions for storage are described in Sections 80.301(a) through 80.301(aa). Specific provisions for each hazard category are described in Sections 80.302 through 80.315. Subsections in each hazard category address:

a)   Indoor storage
b)   Exterior storage
c)   Special provisions

For materials specifically addressed elsewhere in the code, cross references are provided for the reader's convenience:

Explosives	Article 77
Compressed Gases	Article 74
Cryogenic Fluids	Article 75
Flammable/Combustible Liquids	Article 79

Division IV - Dispensing, Use and Handling. This division applies to the dispensing, use and handling of hazardous materials in excess of exempt amounts (Tables 80.402A and 80.402B). The division is further divided into three major sections:

a)   General
b)   Dispensing and Use
c)   Handling

When addressing dispensing and use, consideration is given to location (interior or exterior) and condition of that dispensing and use (open or closed container).

The "Handling" section addresses on-site transportation of hazardous materials within a facility including:

a)   Container types and sizes
b)   Cart and truck specifications/methods
c)   Emergency alarms

1.4.5   Article 80-1 Silane: There is a <u>proposed</u> ordinance which, if adopted, would appear in the 1994 edition of the UFC to deal with the specific hazards associated with the use and storage of silane gas (SiH_4). While it may be inappropriate to discuss specific proposed rules in this book, it is worth mentioning that 80-1 would apply to installations using silane or gas mixtures with more than 2% silane in quantities of more than 100 cubic feet (in sprinklered buildings in gas cabinets).

1.5   OCCUPANCY CLASSIFICATIONS OF PROPOSED FACILITIES

The UBC and UFC codes are used to help the owner and designer define the category of occupancy to be used as a basis for the design, construction and operation of a facility. The various occupancy groups are:

A	-	Assembly	I	-	Institutional
B	-	Business	M	-	Miscellaneous
E	-	Educational	R	-	Residential
H	-	**Hazardous**			

Within each of the occupancy groups are sub-categories called Divisions, which divide the larger group into specific use categories. Under the Group H (Hazardous) occupancy category, there are seven divisions which relate to specific uses. The following descriptions are <u>paraphrased</u> and abbreviated from the UBC (refer to Sec. 901(a)):

Division 1.   Storage, handling and use of hazardous highly-flammable explosive material, in excess of the exempt amounts listed in Table 9-A. This is an occupancy which presents a high explosion hazard (not generally associated with semiconductor facilities).

Division 2.   Storage, handling and use of flammable or combustible liquids, pyrophoric gases and oxidizers in excess of the exempt amounts listed in Table 9-A. This is an occupancy which presents a moderate explosion hazard or hazard from accelerated burning.

Division 3.   Storage or use (low pressure) of flammable and combustible liquids, oxidizers or water reactives. This is an occupancy which presents a high fire or physical hazard to personnel or construction.

Division 4.   Repair garages.

Division 5.   Aircraft repair hangars.

Division 6.   Semiconductor fabrication facilities and comparable research and development areas where hazardous production materials (HPMs) are used, and the aggregate amount exceeds that listed in Tables 9-A or 9-B. Design and construct in accordance with Section 911.

Division 7.   Storage or use of health hazard materials, such as corrosives, toxic or highly toxic irritants or sensitizers in excess of the quantities listed in Table 9-B.

While it is sometimes customary to speak of the "H-6" code requirements with respect to semiconductor facilities, it is important to realize that this is only a generic term used to indicate a body of codes related to such facilities. The "H-6" provisions are specific to Chapter 9 of the Uniform Building Code (UBC) and cannot be taken out of the context of the entire package of hazardous facility codes. This book may speak of the "H-6" code for convenience, however, this means all related codes as well.

For the purposes of typical semiconductor facilities, the divisions which are of particular interest are Division 2, Division 3, Division 6 and Division 7 of the Uniform Building Code and Articles 51, 79 and 80 of the Uniform Fire Code. These divisions will be discussed in detail in the following chapters of this book.

1.6   RETROFIT AND RENOVATION

Often, a concern to the user and owner is the retrofit or modernization of an existing facility. Many industrial facilities currently in use need to be upgraded to allow manufacturing of more sophisticated products. When a retrofit is contemplated, the impact of new code provisions on the construction and operation of that facility must be addressed. The owner, facility staff and code regulators must agree upon a reasonable level of code compliance when full compliance with the codes would be prohibitively expensive or otherwise not feasible.

Owners often believe their facilities are "grandfathered" because they were constructed prior to adoption of the current codes. While there may be a case for acceptance of certain specific non-conformities in a facility which is to remain unchanged, this cannot be taken for granted. A thorough assessment of the facility should be accomplished periodically to ensure its continued safety and compliance with the intent of the law (codes).

Semiconductor and many other manufacturing facilities that were in existence and operation prior to the current codes were generally classified as Group B, Division 2 (B-2) occupancies in lieu of Group H. The UBC administrative provisions allow an area to remain a B-2 occupancy unless changes are made to the use, structure, or support systems, with the following provisions:

> **UBC 104(a) Application to Existing Buildings and Structures**
> "Buildings and structures to which additions, alternations or repairs are made shall comply with all the requirements of this code for new facilities, except as specifically provided in this section. . . ."

UBC 104(b) Application to Existing Buildings and Structures - Additions, Alterations or Repairs

"Additions, alterations or repairs may be made to any building or structure without requiring the existing building or structure to comply with all the requirements of this code, provided the addition, alteration or repair conforms to that required for a new building or structure. . . . nor shall such additions. . . . cause the existing building or structure to become unsafe. An unsafe condition. . . . will not provide adequate egress. . . . will create a fire hazard. . . . or otherwise create conditions dangerous to human life. . . ."

UBC 104(c) Application to Existing Buildings and Structures - Existing Installations

"Buildings in existence at the time of the adoption of this code may have their existing use or occupancy continued, if such use or occupancy was legal at the time of the adoption of this code, provided such continued use is not dangerous to life. Any change in the use or occupancy of any existing building or structure shall comply with the provisions of Sections 308 and 502 of this code."

Discussion:

The statement in UBC 104(c), "provided such continued use is not dangerous to life," has serious implications for design professionals, building officials and building owners alike. The method for evaluating an existing structure's condition and level of safety must be determined on a case-by-case basis.

UBC 308(a) (Permits and Inspections) Certificate of Occupancy - Use and Occupancy

"No building or structure shall be used or occupied, and no change in the existing occupancy classification of a building or structure or portion thereof shall be made until the building official has issued a Certificate of Occupancy therefore as provided herein. . . .

> Issuance of a Certificate of Occupancy shall not be construed as an approval of a violation of the provisions of this code. . . . "

Discussion:

You must obtain a Certificate of Occupancy for changes; this means you must obtain a permit. <u>Also</u>, the building official is not responsible for violations of the codes, even after issuing a Certificate of Occupancy; the building owner is.

UBC 502 Requirements Based on Occupancy - Change in Use

> "No change shall be made in the character of occupancies or use of any building which would place the building in a different division of the same group of occupancy or in a different group of occupancies, unless such building is made to comply with the requirements of this code for such divisions or group of occupancy. . . .
> No change in the character of occupancy of a building shall be made without a Certificate of Occupancy, as required in Section 308 of this code. . . . "

Discussion:

The amount of change in a facility which will be tolerated by your local building official, before a "change in the character or use" is considered, will vary widely. Such a change may require the facility be re-classified from B-2 to H-6, H-7 (or other H division), if you use HPMs in excess of the amounts listed in UBC Tables 9-A or 9-B.

The really difficult issue to address is, what <u>should be</u> done, in addition to what <u>must be</u> done? It is generally accepted that the <u>code intent</u> is that any system which is modified, as a result of the project, must be brought into compliance.

An example of a situation where code regulators are taking a pro-active role in life safety is the promulgation of Toxic Gas Ordinances in several jurisdictions. In these ordinances, the owner of a facility using toxic gases must comply with the new regulations (integrated with UFC Article 80), whether or not the facility is modified. In these jurisdictions, the concept of a "grandfathered" facility, at least as it applies to toxic gases, is eliminated. An excerpt from the City of Palo Alto, California Ordinance (Number 3952) is included here as an example:

> ". . . . WHEREAS, the 1987-88 Santa Clara County Grand Jury found that the use of toxic gases in industry posed a threat to the health and safety of local residents, and recommended that the Intergovernmental Council ("IGC") 'develop an ordinance acceptable to industry, citizen groups, and local government'; and . . . . "

The ordinance goes on to stipulate the exempt amounts of hazardous/toxic gases, which, if exceeded, will cause a facility to fall within the jurisdiction of the ordinance. If the facility is within that jurisdiction then (underlining added):

> ". . . . 90.101 Application. (a) This article applies to all new and existing facilities where regulated materials subject to this article are present in concentrations which exceed the Level of Concern as determined in accordance with this article. . . . "

> ". . . . 90.201(e) Compliance.
> 1. Notwithstanding Section 1.103 of the Fire Code, persons responsible for any facility lawfully in existence on April 16, 1990 which is not in compliance with the provisions of this article shall submit a compliance plan to the

<u>Fire Chief no later than April 16, 1991</u>. For purposes of this section, the term "lawfully in existence" includes, but is not limited to, those facilities for which . . . .

2. Persons responsible for facilities lawfully in existence on April 16, 1990 shall cause their facilities <u>to be in full compliance with this article not later than April 16, 1993</u>. The Fire Chief may extend this time period at the request of a ...."

In other words, the new <u>or existing</u> facility is required to comply with the ordinance (essentially invoking the requirements of UFC Article 80) by virtue of the Grand Jury having found the use of toxic gases to be "potentially dangerous." This finding of the Grand Jury took away the ability of an owner or code official to claim the facility was "grandfathered" under the provisions of UBC 104(c).

1.7 UNDERSTANDING CODE INTENT AND PHILOSOPHY

It is crucial to understand the <u>intent</u> and philosophy of the various codes in order to properly apply them to specific situations. When a retrofit or addition is contemplated for an existing non-conforming facility, for example, the owner/design team must address the question of how far to go in bringing the facility up to the new code requirements. The issues go beyond the matter of what is legally required, for this is subject to interpretation by the building official and negotiation by the owner.

During the design of any new project or retrofit of an existing facility, it is important to establish a working relationship with the local code enforcement agencies to ensure they understand the goals of the project and "buy into" the proposed design and code compliance strategy early on.

There are many ways of satisfying the intent of the code on any given issue; thus, a single method of compliance will not be appropriate in every situation. Many times, alternative methods are acceptable to building officials, when strict compliance with the code is not practical or is inappropriate. In situations where alternative compliance means are proposed, evaluations are done on a case-by-case basis by the building official.

UBC 105 Alternate Materials and Methods of Construction

"The provisions of this code are not intended to prevent the use of any material or method of construction not specifically prescribed by this code, provided any alternate has been approved and its use authorized by the building official.

The building official may approve any such alternate, provided he finds that the proposed design is satisfactory and complies...and is equivalent...in suitability, strength, effectiveness, fire resistance, durability, safety and sanitation."

UBC 106 Modifications

"When there are practical difficulties involved in carrying out the provisions of this code, the building official may grant modifications for individual cases. The building official shall first find that a special, individual reason makes the strict letter of this code impractical and...that the modification is conformance with the intent and purpose of this code..."

UFC 2.301(a) (Special Procedures) Alternate Materials and Methods - Practical Difficulties

"The chief is authorized to modify any of the provisions of this code upon application in writing by the owner, a lessee or a duly authorized representative where there are practical difficulties in the way of carrying out the provisions of the code, provided that the spirit of the code shall be complied with, public safety secured and substantial justice done. . ."

UFC 2.301(b) (Special Procedures) Alternate Materials

"The chief, on notice to the building official, is authorized to approve alternate material or method, provided that the chief finds...at least equivalent...quality, strength, effectiveness, fire resistance, durability and safety. . ."

1.8   INSURANCE CARRIER REQUIREMENTS

The fact that all codes represent <u>minimum</u> acceptable standards may be emphasized by your insurance carrier. While some companies are self insured, most utilize an outside carrier.

Insurance carriers such as Factory Mutual (FM) or Industrial Risk Insurers (IRI) generally will have specific requirements which exceed the code standards. In order to maintain coverage, or perhaps to obtain a lower premium, a facility owner may be required to comply with the carrier's standards, in addition to those of the building official.   Areas which frequently come under insurance carrier scrutiny include:

- Exiting
- Fire Separations
- Construction Materials
- Fire Suppression Systems
- Ventilations Rates

It is good practice to bring the insurance carrier into the design process early in the game, to ensure the optimum solution and avoid delays or excessive premium cost.

An example of such requirements is the Factory Mutual Engineering Corporation Loss Data Sheet 7-7 entitled "Semiconductor Fabrication Facilities" dated October, 1991. This document provides prescriptive requirements, engineering guidance and case histories related to actual losses associated with wafer processing operations.

1.9   DESIGNER / BUILDING OFFICIAL / OWNER RELATIONSHIP AND LIABILITIES

The relationship between the designer, building owner and building official throughout the design, construction and operation of any facility is extremely important.

The **Design Professional** is responsible for proposing methods of compliance with the intent of the various codes applicable to the project. The design professional must be able to interpret the philosophy and underlying concepts of the various codes and put each code requirement into the context of the overall project. The designer must be able to integrate the life safety aspects of the code into a practical design which meets the owner's needs.

The **Building Official** is responsible for reviewing and verifying the design with respect to the applicable codes. The building official is not charged with insuring compliance methods. The building official may make recommendations or advise the designer or owner; however, in no case will the building official assume responsibility for issues of design. It is important to realize that rulings issued by the building official and interpretations of the building official do not absolve either the designer or the owner from the responsibility for compliance with the code. In other words, an erroneous ruling of a building official does not remove the burden of construction and operation of a safe facility from the designer and owner.

In many cases, code compliance strategies are interpreted by the local building official based on his understanding of the intent of the code. In situations where the owner or designer believes that the building official has misinterpreted the requirements and intent of the code, the code agencies have provisions for providing "second opinions" regarding such interpretations. The national code agencies frequently publish interpretations and memoranda pertaining to specific code issues. You may request that the local building official obtain an interpretation of a specific application of the code from ICBO (or BOCA) to ensure the intent is met.

In addition, UBC Section 204 and UFC Section 2.303 provide for an appeal process if a particular ruling is in dispute and cannot be otherwise resolved:

> ### UBC Section 204(a) Organization and Enforcement - Board of Appeals
> "In order to hear and decide appeals of orders, decisions or determinations made by the building official relative to the application and interpretation of this code, there shall be and is hereby created a Board of Appeals consisting of members who are qualified by experience and training to pass upon matters pertaining to building construction..."

The **Owner** of a facility is the individual or company ultimately responsible for the safety and welfare of the employees and public using the facility. Thus, the ultimate burden of complying with the intent of all codes rests with the building owner. While the owner may have recourse against a designer in the event of an accident or other incident, the owner will bear the burden of proof and ultimate liability. The code standards provide minimum requirements and do not relieve the owner from the need to build and operate a safe environment. In many cases, the codes lag behind the industry understanding of potential hazards and abatement or mitigation procedures. Therefore, an owner cannot rely upon code compliance as a defense, if a catastrophic event occurs.

The owner and design team should incorporate into the facility as many techniques available for maintenance of life safety as are feasible. The building owner should consider all potential incident scenarios and evaluate the credible events which might occur during the life of the facility. Such evaluation will help ensure that the facility is safe for the occupants and help protect the owner's investment.

(Refer to **Figure 1.3** *for an illustration of the above relationship.)*

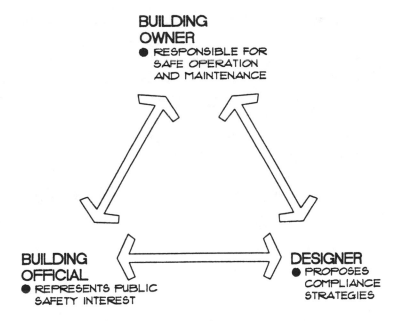

Figure 1.3 RELATIONSHIP DIAGRAM Illustrates Respective Roles of Each Party in a Facility Project.

1.10   INTEGRATED APPROACH

We have stated above, and it bears repeating, that the UBC, the UFC and the related package of codes adopted by a local jurisdiction must be used in their entirety and in concert with each other.   Many designers and building owners go astray by taking one code reference out of context with the body of the codes, and, therefore, misunderstand and misinterpret the intent.   In all cases, the most stringent requirement of related codes must be applied to a specific design or operational criteria.   Careful evaluation of the various codes, as presented in later chapters of this book, will illustrate overlap in their requirements.   In some cases, one code will override another code by virtue of its having more stringent requirements.

A WORD OF CAUTION: The owner or designer of a project must not use this book or the opinions expressed or implied herein as the basis for the design or operation of a facility. Each project is unique in its requirements, and the requirements of each building official or jurisdiction will be unique. Therefore, the following information is provided to outline general procedures and guidelines; these procedures and guidelines are not all-inclusive, nor are they sufficient for the design of any given project.

As with any law, ignorance of a code requirement is not a defense for noncompliance.

SUMMARY

HISTORY OF THE HAZARDOUS OCCUPANCY CODES (1.1)

◆   Codes are ordinances (laws) adopted by the local government jurisdiction to establish minimum guidelines for construction and/or operation of a facility to protect life and property.

◆   Prior to 1985, most wafer fabs and similar industrial occupancies were designed under the provisions of the Uniform Building Code (UBC) Group B - Division 2.

◆   In 1985, the Uniform Building Code and the Uniform Fire Code (UFC) first published Section 911 and Articles 51 and 80, respectively. These codes specifically address building and operating requirements for semiconductor fabrication and other facilities utilizing hazardous materials.

◆   This book is based upon the 1991 editions of the UBC, UFC and related "model" codes.

OVERVIEW OF APPLICABLE CODES (1.2)

◆   Today, the basic "model" codes which affect semiconductor fabrication and other hazardous industrial facilities are numerous.

◆   In addition, corporate safety directives and insurance underwriter requirements must be considered.

OVERVIEW OF UNIFORM BUILDING CODE (1.3)

◆   Without diminishing the need to consider all requirements of the code, the primary chapters in the UBC which are of importance in the design and construction of a hazardous occupancy include:

Chapter 5   Occupancy Classification and Requirements

Chapter 9   Hazardous Occupancies (Group H)
Chapter 17   Types of Building Construction Classifications and Requirements
Chapter 33   Exit Requirements
Chapter 43   Fire-Resistive Construction Requirements

OVERVIEW OF UNIFORM FIRE CODE (1.4)

♦   The articles of greatest interest to our discussion of hazardous occupancies in the Uniform Fire Code are:

Article 51   Semiconductor Facilities
Article 79   Flammable and Combustible Liquids
Article 80   Hazardous Materials

OCCUPANCY CLASSIFICATIONS OF PROPOSED FACILITIES (1.5)

♦   The UBC and UFC codes are used to help the owner and designer define the category of occupancy to be used as a basis for the design, construction and operation of a facility.

A	-	Assembly	I	-	Institutional
B	-	Business	M	-	Miscellaneous
E	-	Educational	R	-	Residential
H	-	**Hazardous**			

♦   When the facility uses or stores hazardous chemicals or gases in sufficient quantity, it must be considered a Hazardous, Group H occupancy.

♦   The UBC and UFC divides hazardous materials into two hazard categories which are directly correlated to Tables 9-A and 9-B of the UBC:

■   Physical Hazards
■   Health Hazards

♦   Within each of the occupancy groups are sub-categories called
         Divisions.   Sec. 901(a) of the UBC provides definitions of Group H
         facilities.

RETROFIT AND RENOVATION (1.6)

♦   When a retrofit is contemplated, the impact of new code provisions on
         the construction and operation of that facility must be addressed. The
         owner, facility staff and code regulators must agree upon a reasonable
         level of code compliance.   The really difficult issue to address is:
         what should be done, in addition to what must be done? In that the
         implications of the code requirements represent significant
         expenditures with the potential for interruption of production, it may
         be desirable to accomplish the work in a phased manner.

♦   Owners often believe their facilities are "grandfathered." While there
         may be a case for acceptance of certain specific non-conformities
         (with current codes) in a facility which is to remain unchanged, this
         cannot be taken for granted. Buildings in existence at the time of the
         adoption of this code may have their existing use or occupancy
         continued, if such use or occupancy was legal at the time of the
         adoption of this code, provided such continued use is not dangerous to
         life. (UBC 104(c))

♦   An example of a situation where code regulators are taking a
         pro-active role in life safety is the promulgation of Toxic Gas
         Ordinances in several jurisdictions. In these ordinances, the owner of
         a facility using toxic gases must comply with the new regulations
         (integrated with UFC Article 80), whether or not the facility is
         modified.

DESIGNER / BUILDING OFFICIAL / OWNER RELATIONSHIP AND LIABILITIES (1.9)

♦   The relationship between the designer, building owner and building official throughout the design, construction and operation of any facility is extremely important.

■   The **Design Professional** is responsible for proposing methods of compliance with the intent of the various codes applicable to the project.   The designer must be able to integrate the life safety aspects of the code into a practical facility which meets the owner's needs.

■   The **Building Official** is responsible for reviewing and verifying the design with respect to the applicable codes. The building official is not charged with insuring compliance methods.

■   The **Owner** of a facility is the individual or company ultimately responsible for the safety and welfare of the employees and public using the facility.   Thus, the ultimate burden of complying with the intent of all codes rests with the building owner.

♦   The code standards provide minimum requirements and do not relieve the owner from the need to build and operate a safe environment. The owner and design team should incorporate into the facility as many techniques available for maintenance of life safety as are feasible.   Current industry standards and knowledge of safe practices frequently are ahead of codes and regulations.

INTEGRATED APPROACH (1.10)

♦   The UBC, UFC, NFPA and the related package of codes adopted by a local jurisdiction must be used in their entirety and in concert with each other.

♦   In all cases, the most stringent requirement of related codes must be applied to a specific design or operational criteria.

2

Architectural Issues

2.1   GENERAL

The "architectural" issues dealt with in the model codes cover a wide range of subjects related to life safety and the preservation of property. While referred to as "architectural issues" in this book for the sake of organization, this chapter will deal with the fundamental organization of a hazardous occupancy and its context with the rest of a facility.

Hazardous occupancies rarely stand alone; they typically comprise one portion of a factory or institution. "H" occupancies require other less dangerous occupancies such as offices, warehouses, assembly areas, etc. for support.

The issues we will deal with in this chapter will establish the ground rules for the design, construction and operation of the project. These issues relate to new facilities being contemplated by an owner, as well as existing facilities constructed under codes established when less awareness existed of the hazards inherent in occupancies using chemicals.

We will deal with the following fundamental principles:

Retrofit & Renovation Criteria   Exiting from the Occupancy
Occupancy Classification   Exit vs. Service (HPM) Corridors
Separation of Occupancies   Chemical Storage & Use Facilities
Allowable Area of Occupancy

2.2   RENOVATION

Much of our discussion concerning the issues related to design of facilities employing hazardous materials will be geared toward the renovation and retrofit of existing semiconductor wafer fabrication facilities and their ancillary occupancies.   Although we will discuss retrofit in the context of a wafer fab, the issues apply to any hazardous occupancies. Thus, the use of a wafer fab in the example is intended to make the discussion less abstract.

Many existing facilities were designed as B-2 occupancies, prior to the advent of UBC occupancy classifications H-1 through H-7. Renovation of these facilities presents serious constraints (both economic and practical) not encountered in new construction.   (*In Chapter 8 of this book we discuss a methodology of prioritizing the issues related to a retrofit and developing a successful project delivery strategy.*)

In order to best determine the method with which to proceed on a renovation project, a thorough understanding of existing conditions in a facility is imperative.   Existing wafer fabs were generally constructed under UBC requirements for B-2 occupancies. UBC B-2 does not address the subject of allowable floor area with respect to chemical usage; prior to 1985, the facilities were allowed to be of unlimited area, provided they were sprinklered and had side yards.

Provisions of UBC Chapter 9 (Hazardous Occupancies) limit the allowable areas of these occupancies with respect to types and quantities of chemicals used.   In addition,   UFC Article 51 (Semiconductor Fabrication Facilities), Article 79 (Flammable and Combustible Liquids) and Article 80 (Hazardous Materials) impose serious design considerations.   In order to proceed wisely, as much information as possible about the existing conditions and use of the facility must be gathered and examined.

The rules governing renovation of an existing facility are outlined in the UBC in Section 502 which states in part:

UBC Sec. 502 Classification of All Buildings by Use - Change in Use

"No change shall be made in the character of occupancies or use of any building which would place the building in a different division of the same group of occupancy or in a different group of occupancies, unless such building is made to comply with the requirements of this code for such division or group of occupancy."

This provision, when taken in context with UBC 104(c) (Existing Installations) as discussed in Chapter 1, may be construed to mean that if the use of a building is not changed, the building does not need to be brought up to current code requirements. The issue of bringing an existing facility into compliance with "H" requirements may be triggered when a renovation, addition or change in function is contemplated. On the other hand, the statement in UBC Section 104(c) "...provided such continued use is not dangerous to life..." must be seriously considered by all associated with the facility. With the application of good judgment and a technical audit, you may recognize that many facilities are not as safe as they might and should be.

The extent of a renovation, addition, or change in function, which may be tolerated before the entire building needs to be brought into compliance, is a matter of the interpretation of the local building official and/or fire marshal. In some jurisdictions, interpretation of the intent of the codes concerning change in "character" or "use" of a building is very strict. As examples, changing a piece of production equipment, or changing the chemicals used in the facility, may be grounds for requiring code upgrade. It is important for the owner and designer of a facility to have a clear understanding of the building officials' stand on this issue, early in the development of a project.

Most building officials understand the financial burden associated with compliance with hazardous occupancy requirements, therefore, the owner may be able to <u>negotiate a phased compliance strategy</u>.

In those jurisdictions where a Toxic Gas (or similar) Ordinance has been enacted, certain aspects of the facility related to storage and use of the HPM gases may need to be brought into compliance with current codes, whether or not any other renovation or changes occur. (*Refer to further discussion in Chapter 3.*)

2.3   OCCUPANCY CLASSIFICATION

One of the first decisions which must be made during the design process concerns the correct occupancy classification(s) for the proposed facility. A summary of the quantity, location and type of hazardous materials is required in order to make the correct occupancy classifications. Classification of the chemicals used or stored in the facility must take into account all aspects of their potential hazard -- such as toxicity, flammability, health hazard, explosivity, etc. -- to determine the true nature of the occupancy. (We wish to point out that many experts on chemicals have differing opinions as to how to classify certain materials. The chemicals which most often fall into the "gray" areas are hazardous gases which may be toxic, flammable, pyrophoric or a combination of these hazards, depending on their state. Your own safety and environmental experts may have opinions on HPM classification which are more stringent than the code agencies' opinions, particularly with regard to toxicity; therefore, your design and operation techniques may be more stringent.)

In the event a review of the hazardous material inventory indicates quantities above the limits (exempt amounts) set in UBC Tables 9-A or 9-B within a given "control area," an "H" occupancy classification is mandated (*see* **Tables 2.1 and 2.2** *for excerpts of Tables 9-A and 9-B*). Note that the tables permit increases in the "exempt" amounts for certain chemicals, when the facility is sprinklered <u>and</u> when the materials are stored in approved storage cabinets.

TABLE NO. 9-A—EXEMPT AMOUNTS OF HAZARDOUS MATERIALS, LIQUIDS
AND CHEMICALS PRESENTING A PHYSICAL HAZARD
BASIC QUANTITIES PER CONTROL AREA[1]
When two units are given, values within parentheses are in cubic feet (Cu. Ft.) or pounds (Lbs.)

MATERIAL	CLASS	STORAGE[2] Solid Lbs.[3] (Cu. Ft.)	Liquid Gallons[3] (Lbs.)	Gas Cu. Ft.	USE[2]—CLOSED SYSTEMS Solid Lbs. (Cu. Ft.)	Liquid Gallons (Lbs.)	Gas Cu. Ft.	USE[2]—OPEN SYSTEMS Solid Lbs. (Cu. Ft.)	Liquid Gallons (Lbs.)
1.1 Combustible liquid[4,5,6,8,9,10]	II	N.A.	120[7]	N.A.	N.A.	120	N.A.	N.A.	30
	III-A	N.A.	330[7]	N.A.	N.A.	330	N.A.	N.A.	80
	III-B	N.A.	13,200[7,11]	N.A.	N.A.	13,200[11]	N.A.	N.A.	3,300[11]
1.2 Combustible dust lbs./1000 cu. ft.[12]		1	N.A.	N.A.	1	N.A.	N.A.	1	N.A.
1.3 Combustible fiber (loose)		(100)	N.A.	N.A.	(100)	N.A.	N.A.	(20)	N.A.
(baled)		(1,000)	N.A.	N.A.	(1,000)	N.A.	N.A.	(200)	N.A.
1.4 Cryogenic, flammable or oxidizing			45	N.A.	N.A.	45	N.A.	N.A.	10
2.1 Explosives[13]		1[7,14]	(1)[7,14]	N.A.	1/4	(1/4)	N.A.	1/4	(1/4)
3.1 Flammable solid		125[6,7]	N.A.	N.A.	25[6]	N.A.	N.A.	25[6]	N.A.
3.2 Flammable gas (gaseous)		N.A.	N.A.	750[6,7]	N.A.	N.A.	750[6,7]	N.A.	N.A.
(liquefied)		N.A.	15[7]	N.A.	N.A.	15[6,7]	N.A.	N.A.	N.A.
3.3 Flammable liquid[4,5,6,8,9,10]	I-A	N.A.	30[7]	N.A.	N.A.	30	N.A.	N.A.	10
	I-B	N.A.	60[7]	N.A.	N.A.	60	N.A.	N.A.	15
	I-C	N.A.	90[7]	N.A.	N.A.	90	N.A.	N.A.	20
Combination I-A, I-B, I-C		N.A.	120[7]	N.A.	N.A.	120	N.A.	N.A.	30
4.1 Organic peroxide, unclassified detonatable		1[7,13]	(1)[7,13]	N.A.	1/4[13]	(1/4)[13]	N.A.	1/4[13]	(1/4)[13]
4.2 Organic peroxide	I	5[6,7]	(5)[6,7]	N.A.	(1)[6]	(1)[6]	N.A.	1[6]	1[6]
	II	50[6,7]	(50)[6]	N.A.	50[6]	(50)[6]	N.A.	10[6]	(10)[6]
	III	125[6,7]	(125)[6,7]	N.A.	125[6]	(125)[6]	N.A.	25[6]	(25)[6]
	IV	500	(500)	N.A.	500[6]	(500)	N..A.	100	(100)
	V	N.L.	N.L.	N.A.	N.L.	N.L.	N.A.	N.L.	N.L.
4.3 Oxidizer	4	1[7,13]	(1)[7,13]	N.A.	1/4[13]	(1/4)[13]	N.A.	1/4[13]	(1/4)[13]
	3[16]	10[6,7]	(10)[6,7]	N.A.	2[6]	(2)[6]	N.A.	2[6]	(2)[6]
	2	250[6,7]	(250)[6,7]	N.A.	250[6]	(250)[6]	N.A.	50[6]	(50)[6]
	1	4,000[6,7]	(4,000)[6,7]	N.A.	4,000[6]	(4,000)[6]	N.A.	1,000[6]	(1,000)[6]
4.4 Oxidizer—Gas (gaseous)[6,7]		N.A.	N.A.	1,500	N.A.	N.A.	1,500	N.A.	N.A.
(liquefied)[6,7]		N.A.	15	N.A.	N.A.	15	N.A.	N.A.	N.A.
5.1 Pyrophoric		4[7,13]	(4)[7,13]	50[7,13]	1[13]	(1)[13]	10[7,13]	0	0
6.1 Unstable (reactive)	4	1[7,13]	(1)[7,13]	10[7,13]	1/4[13]	(1/4)[13]	2[7,13]	1/4[13]	(1/4)[13]
	3	5[6,7]	(5)[6,7]	50[6,7]	1[6]	(1)[6]	10[4,5]	1[6]	(1)[6]
	2	50[6,7]	(50)[6,7]	250[6,7]	50[6]	(50)[6]	250[4,5]	10[6]	(10)[6]
	1	N.L.	N.L.	750[6,7]	N.L.	N.L.	N.L.	N.L.	(25)[6]
7.1 Water (reactive)	3	5[6,7]	(5)[6,7]	N.A.	5[6]	(5)[6]	N.A.	1[6]	(1)[6]
	2	50[6,7]	(50)[6,7]	N.A.	50[6]	(50)[6]	N.A.	10[6]	(10)[6]
	1	125[7,11]	(125)[7,11]	N.A.	125[11]	(125)[7,11]	N.A.	25[11]	(25)[11]

N.A. = Not applicable.   N.L. = Not limited.

[1] Control area is a space bounded by not less than a one-hour fire-resistive occupancy separation within which the exempted amounts of hazardous materials may be stored, dispensed, handled or used. The number of control areas within a building used for retail and wholesale stores shall not exceed two. The number of control areas in buildings with other uses shall not exceed four.

[2] The aggregate quantity in use and storage shall not exceed the quantity listed for storage.

[3] The aggregate quantity of nonflammable solid and noncombustible liquid hazardous materials within a single control area of Group B, Division 2 Occupancies used for retail sales may exceed the exempt amounts when such areas are in compliance with the Fire Code.

[4] The quantities of alcoholic beverages in retail sales uses are unlimited provided the liquids are packaged in individual containers not exceeding four liters.
The quantities of medicines, foodstuffs and cosmetics containing not more than 50 percent of volume of water-miscible liquids and with the remainder of the solutions not being flammable in retail sales or storage occupancies are unlimited when packaged in individual containers not exceeding four liters.

[5] For aerosols, see the Fire Code.

[6] Quantities may be increased 100 percent in sprinklered buildings. When Footnote No. 7 also applies, the increase for both footnotes may be applied.

[7] Quantities may be increased 100 percent when stored in approved storage cabinets, gas cabinets, fume hoods, exhaust enclosures or safety cans as specified in the Fire Code. When Footnote No. 6 also applies, the increase for both footnotes may be applied.

[8] For storage and use of flammable and combustible liquids in Groups A, B, E, I, M and R Occupancies, see Sections 608, 708, 808, 1008, 1104 and 1213.

[9] For wholesale and retail sales use, also see the Fire Code.

[10] Spray application of any quantity of flammable or combustible liquids shall be conducted as set forth in the Fire Code.

[11] The quantities permitted in a sprinklered building are not limited.

[12] A dust explosion potential is considered to exist if 1 pound or more of combustible dust per 1,000 cubic feet of volume is normally in suspension or could be put into suspension in all or a portion of an enclosure or inside pieces of equipment. This also includes combustible dust which accumulates on horizontal surfaces inside buildings or equipment and which could be put into suspension by an accident, sudden force or small explosion.

[13] Permitted in sprinklered buildings only. None is allowed in unsprinklered buildings.

[14] One pound of black sporting powder and 20 pounds of smokeless powder are permitted in sprinklered or unsprinklered buildings.

[15] Containing not more than the exempt amounts of Class I-A, Class I-B or Class I-C flammable liquids.

[16] A maximum quantity of 200 pounds of solid or 20 gallons of liquid Class 3 oxidizers may be permitted in Groups I, M and R Occupancies when such materials are necessary for maintenance purposes or operation of equipment as set forth in the Fire Code.

Table 2.1   UBC EXEMPT AMOUNTS OF PHYSICAL HAZARD MATERIAL (UBC Table 9-A).

TABLE NO. 9-B—EXEMPT AMOUNTS OF HAZARDOUS MATERIALS, LIQUIDS
AND CHEMICALS PRESENTING A HEALTH HAZARD

MAXIMUM QUANTITIES PER CONTROL AREA[1,2]

When two units are given, values within parentheses are in pounds (Lbs.)

MATERIAL	STORAGE[3]			USE[3]—CLOSED SYSTEMS			USE[3]—OPEN SYSTEMS	
	Solid Lbs.[4,5,8]	Liquid Gallons[4,5,8] (Lbs.)	Gas Cu. Ft.[5]	Solid Lbs.[4,5]	Liquid Gallons[4,5] (Lbs.)	Gas Cu. Ft.	Solid Lbs.[4,5]	Liquid Gallons[4,5] (Lbs.)
1. Corrosives	5,000	500	650[6]	5,000	500	650[5,6]	1,000	100
2. Highly Toxics[8]	1	(1)	20[7]	1	(1)	20[7]	1/4	(1/4)
3. Irritants	5,000	500	650[6]	5,000	500	650[5,6]	1,000	100
4. Sensitizers	5,000	500	650[6]	5,000	500 .	650[5,6]	1,000	100
5. Other Health Hazards	5,000	500	650[6]	5,000	500	650[5,6]	1,000	100
6. Toxics	500	(500)	650[6]	500	(500)	20[5,7]	125	(125)

[1]Control area is a space bounded by not less than one-hour fire-resistive occupancy separation within which the exempted amounts of hazardous materials may be stored, dispensed, handled or used. The number of control areas within retail and wholesale stores shall not exceed two and the number of control areas in other uses shall not exceed four.

[2]The quantities of medicines, foodstuffs and cosmetics, containing not more than 50 percent by volume of water-miscible liquids and with the remainder of the solutions not being flammable, in retail sales uses are unlimited when packaged in individual containers not exceeding 4 liters.

[3]The aggregate quantity in use and storage shall not exceed the quantity listed for storage.

[4]The aggregate quantity of nonflammable solid and nonflammable or noncombustible liquid health hazard materials within a single control area of Group B, Division 2 Occupancies used for retail sales may exceed the exempt amounts when such areas are in compliance with the Fire Code.

[5]Quantities may be increased 100 percent when stored in sprinklered buildings. When Footnote No. 6 also applies, the increase for both footnotes may be applied.

[6]Quantities may be increased 100 percent when stored in approved storage cabinets, gas cabinets, fume hoods, exhausted enclosures or safety cans as specified in the Fire Code. When Footnote No. 5 also applies, the increase for both footnotes may be applied.

[7]Permitted only when stored in approved exhausted gas cabinets, exhausted enclosures or fume hoods.

[8]For special provisions, see the Fire Code.

The "control area" is defined in UBC Section 404 as follows:

> **CONTROL AREA** is a space bounded by not less than a one-hour fire resistive <u>occupancy separation</u> within which the exempted amounts of hazardous materials may be stored, dispensed, handled or used.

The control area is a concept provided in both the UBC and UFC to reduce the hazard to adjacent spaces caused by (limited) chemical storage or use within the area. Examination of the quantities "exempted" from the requirements of the codes for an "H" occupancy will reveal that they are quite significant for relatively low hazard materials (such as Class III-B combustible liquids, and corrosives), yet they are very restrictive for high hazard materials (such as Class I-A flammable liquids, pyrophorics or highly toxics). The number of control areas within a retail and wholesale store is limited to two, while up to four control areas are allowed in other types of facilities (Footnote No. 1 of Tables 9-A and 9-B). Thus, through the provision of one-hour occupancy separation walls, a facility may contain significant quantities of hazardous materials and remain within the occupancy classification of its primary use (such as a B-2). *The concept of the control area is further explained in Chapter 3.*

In general, once the exempt amount of hazardous material is exceeded within a control area, an "H" occupancy is mandated on the basis of the hazard or hazards presented by the materials. Section 901 of the UBC and Article 80 of the UFC explain the correct occupancy division(s) with Group H for the hazard(s) associated with the materials. When the hazards presented by the materials fall into more than one division, then the occupancy must meet the requirements for all of the applicable divisions. A common example of this occurs with hazardous gases which are both flammable and toxic; the correct occupancy would be Group H, Divisions 2 and 7.

Once the exempt amount has been exceeded, and the area has been classified a hazardous occupancy, the amount of hazardous material which may be used or stored within the occupancy is limited by the following:

- Maximum area of the occupancy - UBC Table 5-C.
- Detached storage requirement of UBC Table 9-E.
- Storage limitations for certain hazardous materials in UFC Article 80.
- Maximum quantities for a "single fabrication area" (H-6) as stipulated by UFC Article 51.

In the case of an H-6 occupancy, Tables 9-A and 9-B of the UBC must be used in conjunction with UFC Tables 51.105-A and 51.105-B to determine the maximum permitted quantities of HPMs. The H-6 classification is unusual, in that it deals with an occupancy (semiconductor manufacturing, or similar R & D areas) that exceeds the exempt amounts of hazardous materials in <u>either</u> Table 9-A (Physical Hazards) or Table 9-B (Health Hazards) or both. (*More on this subject later.*)

UFC Article 80.103 requires permits for the storage, dispensing, use or handling of hazardous materials in excess of the quantities specified in UFC 4.108. The amounts of such materials requiring permits are significantly less than the exempt amounts for each (refer to Tables 4.108A, B or C in the UFC).

Mixed occupancies often exist in a given facility due to storage of hazardous materials in the building and/or administrative areas adjacent to the manufacturing or processing functions. In a building where the area of a different occupancy is less than 10 percent of the area of the primary occupancy, a mixed occupancy is not considered to occur; this is considered a "minor accessory use" condition. However, when a different occupancy in such a situation exceeds 10 percent of the area of the primary occupancy, UBC considers the building a mixed occupancy and requires the summation of the ratio of actual area to allowable area for each occupancy to add up to one or less. This can be quite restrictive when working within certain construction types or with chemical or gas storage occupancies.

UBC 503(a) Classification of All Buildings By Use - Mixed Occupancy

"When a building is used for more than one occupancy purpose, each part of the building comprising a distinct "occupancy," as described in Chapters 5 through 12, shall be separated from any other occupancy as specified in Section 503(d) . . . An occupancy shall not be located above the story or height set forth in Table 5-D, except as provided in Section 507 . . . When a mixed occupancy building contains a Group H, Division 6 Occupancy, the portion containing the Group H, Division 6 Occupancy shall not exceed three stories or 55 feet in height."

UBC 503(d) (Classification of All Buildings By Use) Mixed Occupancy - Fire Ratings for Occupancy Separations

"Occupancy separations shall be provided between the various groups and divisions of occupancies as set forth in Table No. 5-B."

The major use of the building determines the occupancy classification, where the minor accessory uses do not occupy more than 10 percent of the area of any floor. When more than one occupancy occurs within a building, fire-resistive occupancy separations must be constructed between them. Table 5-B and Sections 503(b), (c), and (d) explain clearly the requirements for occupancy separation construction. (*Refer also to* **Figure 2.1**, *for an example of a typical mixed occupancy facility*.)

Note that an H-1 occupancy may <u>not</u> be mixed with any other occupancy.

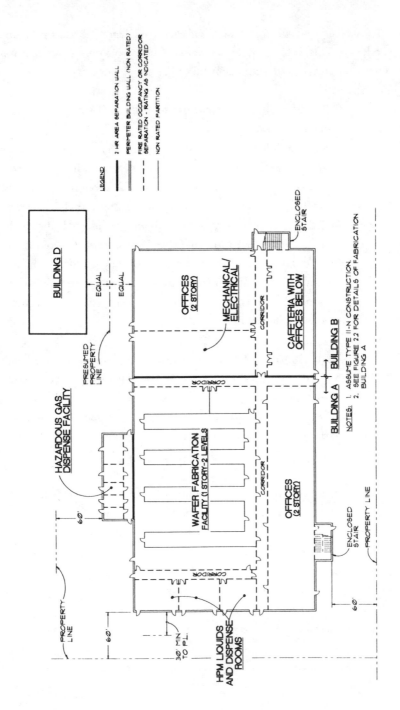

Figure 2.1 PROJECT DIAGRAM illustrates relationship of various buildings and occupancies to one another and to the property lines.

H-7 Occupancies

The 1988 Uniform Building Code identified the category of "H-7" occupancy. H-7 is a classification for facilities which store and use large quantities (above the "exempt" amounts) of materials which are considered "health hazards." Health hazards are those materials such as toxics and corrosives which generally do not represent a danger of fire or explosion, yet are potentially life threatening. Note that the 1993 Edition of the BOCA National Building Code provides for "Use Group H-4," which is similar to UBC H-7.

The use of an H-7 occupancy may be appropriate where there is a need to store or use large quantities of corrosive materials. An H-7 storage facility is often appropriate as an accessory to an H-6. A facility, such as a plating shop, which uses large quantities of corrosives or small quantities of toxics in the process may appropriately be classified "H-7." As an example, Article 51 of the UFC stipulates in Table 51.105-B a maximum of 165 gallons of corrosive liquid in a single H-6 occupancy, whereas, Table 9-B of the UBC lists 500 gallons of corrosives as the exempt amount.

Research laboratories or pilot production facilities may be appropriately classified as H-7 occupancies. This is because quantities of physical hazard chemicals used are generally relatively small; thus, the controlling criteria may be the quantity of highly toxic materials. Under Table 9-B the exempt amount is four pounds stored in a sprinklered building with approved cabinet. It is conceivable that in industry, a pilot production facility or research and development laboratory may be better classified an H-7 Occupancy than an H-6 Occupancy. Each new construction or existing construction retrofit project must be analyzed separately to determine the viability of H-7 Classification and the appropriate architectural, mechanical, electrical and process piping design needs.

While the 1988 UBC identified the H-7 category, few specific construction requirements are identified. Therefore, such issues as ventilation, allowable area, separation from other occupancies, and the like are treated in the various tables and general body of text for H occupancies.

Local code authorities and risk underwriters will have their own interpretations of H-7. These entities must, therefore, be brought into discussions regarding occupancy classification/requirements early in the design and construction process to ensure the success of the project.

2.4   ALLOWABLE AREA AND SEPARATIONS

Facilities which were originally constructed as B-2 occupancy were generally allowed to have unlimited area in a single building if fully sprinklered and surrounded by adequate yards (UBC 506(b)). As such facilities are converted to Group "H" occupancy, severe restrictions on the allowable area may be imposed by the UBC. Buildings rated construction Type I-F.R. are the only facilities allowed unlimited area for an H-3, H-4, H-5, H-6, or H-7 occupancy.

The designer must calculate the maximum allowable area using the correct occupancy and construction type in conjunction with UBC Table 5-C and UBC Sections 505 and 506. This calculated allowable area must then be compared with the existing area in the facility for each occupancy to determine the number and location of area separations which need to be upgraded or constructed. (Area separations are walls and penetrations of the required fire-resistance rating.) This area test must be performed in addition to the tests in UBC and UFC for maximum allowable quantities of HPMs.

UBC 505(a)   (Classification of All Buildings by Use) Allowable Floor Areas - One-Story Areas

"The area of a one-story building shall not exceed the limits set forth in Table No. 5-C, except as provided in Section 506."

UBC 505(b) (Classification of All Buildings by Use) Allowable Floor Areas - Areas of Buildings Over One Story

"The total combined floor area for multi-story buildings may be twice that permitted by Table No. 5-C for one-story buildings, and the floor area of any single story shall not exceed that permitted for a one-story building."

(Refer to **Figures 2.2 and 2.3** *for a graphical presentation of the process of defining and qualifying a facility under the constraints of UBC allowable areas.* **Table 2.3**, *which illustrates the calculation procedure, is explained later.)*

UBC 505(e) Allowable Floor Areas - Basements

"A basement need not be included in the total allowable area, provided such basement does not exceed the area permitted for a one-story building."

Discussion: Many wafer fabrication occupancies have basements used for the support of fabrication operations. If the lowest level is <u>not</u> considered a first-story (*see definitions in Section 403 and 420*), then it does not enter into the calculation of allowable area.

UBC 420 Definitions and Abbreviations - Story

". . . If the finished floor level directly above a usable or unused under-floor space is more than six feet above grade as defined herein for more than 50 percent of the total perimeter or is more than 12 feet above grade as defined herein at any point, such usable or unused under-floor space shall be considered as a story."

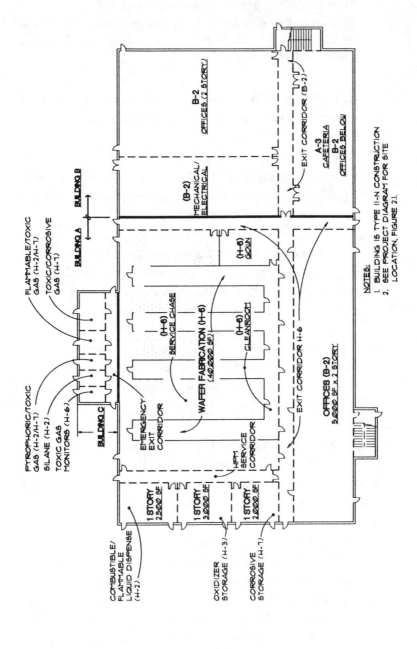

Figure 2.2 OCCUPANCY CLASSIFICATION DIAGRAM illustrates various groups and divisions of occupancies in the project, with corresponding areas for the example on allowable area calculations as shown in **Table 2.3**.

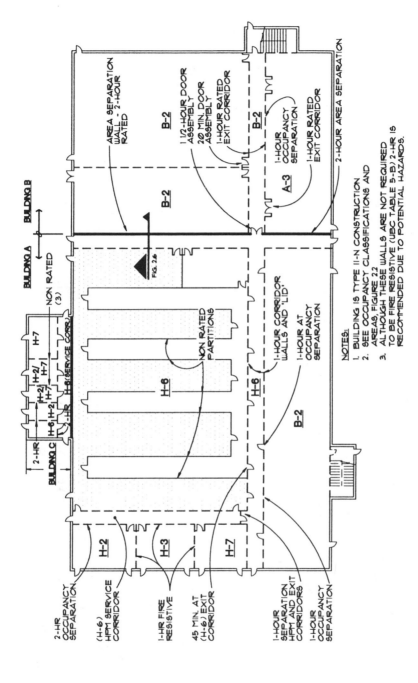

Figure 2.3 SEPARATIONS of the various occupancy types and exitways from one another through the use of fire-resistive construction.

BUILDING "A" AREA ANALYSIS (REFER TO FIGURES 2.2 AND 2.3)

SCENARIO I - ALL OF BUILDING " A " UNSUBDIVIDED

BUILDING TYPE :: II - N

MIXED OCCUPANCY RULE APPLIES BECAUSE B - 2 AREA EXCEEDS 10 % TOTAL AREA

THEREFORE, TEST FOR THE SUMMATION OF AREA RATIOS < 1.0

AREA FUNCTION	UBC DIVISION	ACTUAL AREA SF	BASIC AREA SF	MULTISTORY FACTOR	SEPARATION FACTOR (1)	SPRINKLER FACTOR (2	RESULTANT ALLOWABLE AREA	AREA RATIO
WAFER FABRICATION	H - 6	40,000	12,000	0	100	300	72,000	0.556
SERVICE CORRIDOR	H - 6	INCL. ABV	12,000	0	100	300	72,000	INCL. ABV.
EXIT CORIDOR	H - 6	INCL. ABV	12,000	0	100	300	72,000	INCL. ABV.
GOWNING	H - 6	INCL. ABV	12,000	0	100	300	72,000	INCL. ABV.
OFFICES	B - 2	10,000	12,000	0	100	300	72,000	0.139
COMBUSTIBLE LIQUID STORAGE / DISPENSE	H - 2	2,500	3,700	0	100	0	7,400	0.338
OXIDIZER LIQUID STORAGE	H - 3	3,000	7,500	0	100	0	15,000	0.200
CORROSIVE LIQUID STORAGE	H - 7	2,000	12,000	0	100	300	72,000	0.028
BUILDING " A " TOTALS		57,500						

SUMMATION OF AREA RATIOS = 1.260
PASS MIXED OCCUPANCY TEST ?   NO

CONCLUSION: CANNOT BE ONE BUILDING ! ADD AREA SEPARATION WALL AND SUBDIVIDE INTO TWO BUILDINGS.

FOOTNOTES :(1) ALL SIDE YARDS OF BUILDING " A " ARE AT LEAST 60 FEET THUS, THE MAXIMUM INCREASE OF 100 % PERTAINS (UBC 506 (a) 3.).
(2) NO INCREASE IN AREA ALLOWED FOR SPRINKLERS IN H - 1, H - 2, OR H - 3 OCCUPANCIES (UBC 506 (a) 2.).

Table 2.3 SAMPLE CALCULATION FOR BASIC MIXED AREA OCCUPANCY AS ILLUSTRATED IN FIGURES 2.2 and 2.3.

(Refer to **Figure 2.4** *for a graphical presentation of the rules concerning a basement or first story.)*

UBC 505(f)1   (Requirements Based on Occupancy) Allowable Floor Areas - Area Separation Walls
"Each portion of a building separated by one or more area separation walls which comply with the provisions of this subsection may be considered a separate building.   The extent and location of such area separation walls shall provide a complete separation. . ." (See full text of code)

UBC 506(a)1-3 Classification of All Buildings by Use - Allowable Area Increases
The floor areas specified in Section 505 may be increased by one of the following:

1.   **Separation on two sides** . . . . not to exceed 50 percent.
2.   **Separation on three sides** . . . . not to exceed 100 percent.
3.   **Separation on all sides** . . . . not to exceed 100 percent.
        (See full text of code)

UBC 506(c)1-4 (Classification of All Buildings by Use) Allowable Area Increases - Automatic Sprinkler Systems
The areas specified in Table No. 5-C and Section 505(b) may be tripled in one-story buildings and doubled in buildings of more than one story if the building is provided with an approved automatic sprinkler system throughout.   The area increases permitted in this subsection may be compounded with that specified in paragraphs 1, 2 or 3 of subsection (a) of this section.   The increases permitted in the subsection shall not apply when automatic sprinkler systems are installed under the following provisions:

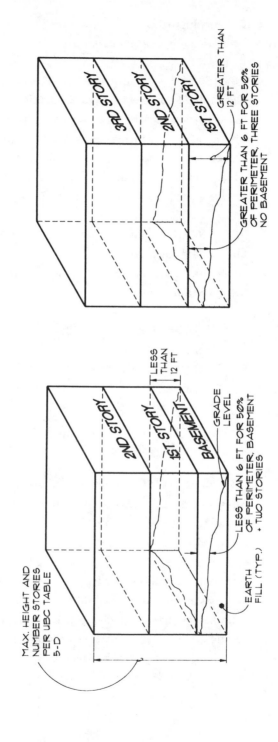

Figure 2.4 BASEMENT OR FIRST STORY. This diagrams the tests for a basement or first story in a building as described in UBC Section 420.

1.   Section 507 for an increase in allowable number of
      stories.

2.   Section 3802(f) for Group H, Division 1, 2 and 3
      Occupancies.

3.   Substitution for one-hour fire-resistive construction
      pursuant to Section 508.

4.   Section 1716, Atria.

(Refer to the sample calculations in **Tables 2.3 and 2.4** *and* **Figures
2.3 and 2.5,** *for illustrations of the application of these provisions.)*

UBC Section 908, Special Hazards, requires that every boiler, central
heating plant or hot water supply boiler be separated from the rest of
the building in which hazardous occupancies occur, by not less than a
two-hour fire-resistive occupancy separation.   This exceeds the
one-hour boiler room separation required in a B occupancy.

Separation of an existing structure into separate buildings (if required
to comply with the maximum allowable area limitations) may be
accomplished by the construction of area separation walls.
Requirements for area separation walls are clearly defined under
Section 505(f). These requirements can severely limit flexibility and
future expansion or renovation. Area separation walls have a major
impact on the mechanical and electrical design (*refer to Chapters 4, 5
and 6*) as well as the architecture; therefore, it is important to
carefully consider the location of any proposed area separation walls
in the code evaluation stage of the project (*see* **Figure 2.6**).

In an existing facility with detached adjacent buildings, the required
area separation is achieved by an exterior fire-resistive rating based on
the distance to the adjacent building, which is treated as though it
were at the property line. UBC Section 504 and Table 5-A clearly
define these requirements.

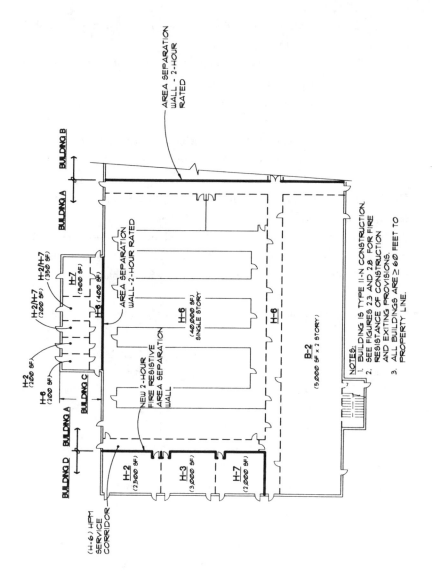

Figure 2.5 FINAL ARRANGEMENT of facility with additional area separation wall at the HPM Liquid Storage area, creating new Building D. Refer to mixed occupancy calculations for each building as shown in **Table 2.4.**

BUILDING "A" AND "D" AREA ANALYSIS (REFER TO FIGURE 2.5)

SCENARIO II - BUILDING "A" SUBDIVIDED AT HPM STORAGE ROOMS (NEW BUILDING "D")

BUILDING TYPE : II - N

MIXED OCCUPANCY RULE APPLIES BECAUSE B 2 AREA EXCEEDS 10 % TOTAL AREA

THEREFORE, TEST FOR THE SUMMATION OF AREA RATIOS < 1.0

AREA FUNCTION - BUILDING "A"	UBC DIVISION	ACTUAL AREA SF	BASIC AREA SF	MULTISTORY FACTOR	SEPARATION FACTOR (1)	SPRINKLER FACTOR (2	RESULTANT ALLOWABLE AREA	AREA RATIO
WAFER FABRICATION	H - 6	40,000	12,000	0	100	300	72,000	0.556
SERVICE CORRIDOR	H - 6	INCL. ABV	12,000	0	100	300	72,000	INCL. ABV.
EXIT CORRIDOR	H - 6	INCL. ABV	12,000	0	100	300	72,000	INCL. ABV.
GOWNING	H - 6	INCL. ABV	12,000	0	100	300	72,000	INCL. ABV.
OFFICES	B - 2	10,000	12,000	0	100	300	72,000	0.139
BUILDING "A" - TOTALS		50,000						

SUMMATION OF AREA RATIOS = 0.694
PASS MIXED OCCUPANCY TEST ? YES

AREA FUNCTION - BUILDING "D"	UBC DIVISION	ACTUAL AREA SF	BASIC AREA SF	MULTISTORY FACTOR	SEPARATION FACTOR (3)	SPRINKLER FACTOR (2	RESULTANT ALLOWABLE AREA	AREA RATIO
COMBUSTIBLE LIQUID STORAGE / DISPENSE	H - 2	2,500	3,700	0	50	0	5,550	0.450
OXIDIZER LIQUID STORAGE	H - 3	3,000	7,500	0	50	0	11,250	0.267
CORROSIVE LIQUID STORAGE	H - 7	2,000	12,000	0	50	300	54,000	0.037
BUILDING "D" - TOTALS		7,500						

SUMMATION OF AREA RATIOS = 0.754
PASS MIXED OCCUPANCY TEST ? YES

CONCLUSION: TWO CODE COMPLIANT BUILDINGS ARE CREATED BY THE AREA SEPARATION WALL AT THE NEW BUILDING "D".

FOOTNOTES : (1) ALL SIDE YARDS OF BUILDING "A" ARE AT LEAST 60 FEET THUS, THE MAXIMUM INCREASE OF 100 % PERTAINS (UBC 506 (a) 3.)
(2) NO INCREASE IN AREA ALLOWED FOR SPRINKLERS IN H - 1, H - 2, OR H - 3 OCCUPANCIES (UBC 506 (c) 2.).
(3) THERE ARE ONLY TWO 60 FOOT SIDEYARDS FOR BUILDING "D". THEREFORE, THE ALLOWABLE INCREASE IS 50 % (UBC 506 (a) 1.).

Table 2.4 SAMPLE CALCULATION FOR REVISED AREA SEPARATIONS AS ILLUSTRATED IN FIGURE 2.5.

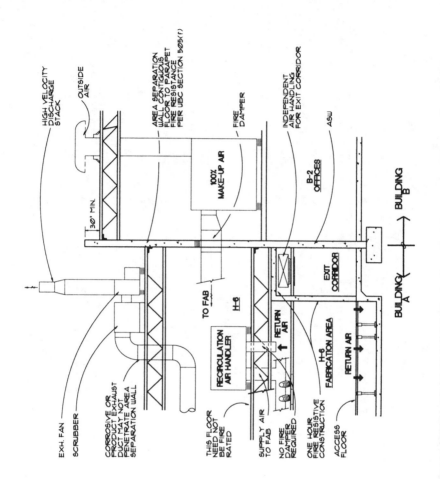

Figure 2.6 SECTION VIEW OF BUILDING illustrating use of area separation wall to isolate two buildings and the separation of air handling systems.

2.5   LOCATION

The location of various occupancies, with respect to the story and height within a building, the proximity to a property line, etc., is generally governed by UBC Chapter 5.

UBC Table 5-A stipulates the fire ratings of exterior walls and openings in walls relative to the distance of such walls from the adjacent property line.

UBC Table 9-C stipulates the rating of fire-resistive construction and protection of openings in exterior walls for each division of Group H occupancy, as a function of distance from the property line.

UBC Table 5-D stipulates the maximum height of a building containing each type of occupancy, as a function of the fire-resistance rating of the structure. In addition, the maximum height in stories of each occupancy is dictated.

UBC Table 9-E stipulates the quantities of explosives, oxidizers, reactives, unstable materials and pyrophoric gases, which, when exceeded, must be stored in a detached building.

UFC 51.105(b) states that occupied levels of fabrication areas shall be located at, or above, the first story. This precludes the occupancy of an "H-6" basement; i.e., if there is a basement, it must be used for equipment only.

2.6   EXITING

Exiting requirements for "H" occupancies are similar to general exiting provisions for other occupancies as set forth in Chapter 33 of the UBC. Section 3319 gives specific requirements for H occupancies. When renovating an area, the designer must first understand the existing exit patterns in the facility as a whole, in order to avoid conflicts between new requirements and existing patterns which must remain. Existing exit corridors, exit passageways, exit travel distances and horizontal exits for adjacent fabrication areas, buildings or other occupancies must be identified and coordinated with the new project exiting plan.

General requirements for exiting:

UBC 3301(b) Exits - Definitions

"EXIT is a continuous and unobstructed means of egress to a public way and shall include intervening aisles, doors, doorways, gates, corridors, exterior exit balconies, ramps, stairways, smoke-proof enclosures, horizontal exits, exit passageways, exit courts and yards."

"EXIT PASSAGEWAY is an enclosed exit connecting a required exit or exit court with a public way."

Generally, exiting requirements are driven by the occupant loading of a facility as dictated by UBC 3302 and 3303.   UBC Table 33-A stipulates the occupant density to be used for various functional occupancies.   In addition, Table 33-A stipulates minimum egress requirements by defining the number of people in a space at which point a minimum of two exits are required.   The following requirements are an excerpt from Table 33-A:

Use	Occupant Load Factor - SF/Person	Number of People Requiring 2 Exits
Manufacturing	200	30
Offices	100	30
Mechanical Equipment Room	300	30
Storage and Stock Rooms	300	30
Warehouses	500	30
"All Others"	100	50

The second story and floors above the second story of a building shall be provided with not less than two exits when the occupant load is 10 or more. Basements and occupied roofs shall be provided with exits as for stories (UBC 3303(a)). Basements shall have not less than two separate exits.

Every story, or portion thereof, having an occupant load of 501 to 1,000, shall not have less than three exits. Every story, or portion thereof, having an occupant load of 1,001 or more, shall not have less than four exits.

The total width of exits, in inches, shall be not less than the total occupant load served multiplied by 0.3 for stairways and 0.2 for other exits, unless more specific requirements are stipulated for a given occupancy. If only two exits are required, they shall be spaced a distance apart equal to not less than one-half the length of the maximum overall diagonal dimension of the building or space.

UBC 3303(d) Exits Required - Distance to Exits

"The maximum distance of travel from any point to an exterior exit door, horizontal exit, exit passageway, or an enclosed stairway in a building . . . shall not exceed....200 feet in a building equipped with automatic sprinkler system . . . These distances may be increased a maximum of 100 feet when the increased travel distance is the last portion of the travel distance and is entirely within a one-hour fire-resistive corridor complying with Section 3305. See . . . Section 3319 for Group H Occupancy travel distances."

UBC 3303(e) Exits Required - Exits Through Adjoining Rooms

"Rooms may have one exit through an adjoining or intervening room which provides a direct, obvious and unobstructed means of travel to an exit corridor, exit enclosure or until egress is provided from the building, provided the total distance of travel does not exceed that permitted...

Exceptions:

2   Rooms with a cumulative occupant load of 10 or less may exit through more than one intervening room."

Corridors serving as required exits for an occupant load of 10 or more shall be not less than 44" inches wide and seven feet high. Corridors shall not be interrupted by intervening rooms and the width shall not be obstructed. Exit corridors shall be arranged to allow travel from any point in two directions to an exit, except for dead-ends not exceeding 20 feet in length. Openings other than doors shall not exceed 25 percent of the area of the corridor wall. Refer to UBC 3305 for specific requirements.

Stairways shall be as specified in UBC 3306 and generally as follows. The minimum width shall be 44" unless occupant load is less than 49, when the width may be 36". Risers shall be not less than four inches or more than seven inches and run shall be not less than 11 inches. Basement stairways which terminate at the same exit as other floors shall have an approved barrier to inhibit occupants of upper floors from continuing on to the basement. Handrails shall be provided on each side. The shaft enclosure for the stairway shall be of fire-resistive construction as required by Table 17-A, generally one- or two-hour, depending on the building type.

Horizontal Exits shall be as specified in UBC 3308 and generally as follows. Horizontal exit is an exit from one building into another building on approximately the same level, or through or around a wall constructed as required for a two-hour occupancy separation and which completely divides a floor into two or more separate areas so as to establish an area of refuge affording safety from fire or smoke coming from the area from which escape is made.

A horizontal exit shall not be the only exit from a portion of a building, and not more than one-half of the total required exits may be horizontal exits. Openings in walls of two-hour fire-resistance shall be not less than 1½-hour resistance. The floor into which the horizontal exit discharges must have adequate capacity to handle the occupant loading of the exit and at least three square feet of clear floor area for each occupant. The area receiving such refuge occupants shall be provided with adequate exits.

Smoke-proof Enclosures shall be as specified in UBC 3310 and generally have the following requirements. The smoke-proof enclosure is required in buildings with occupied floors more than 75 feet above the level of fire department vehicle access and consists of a vestibule and continuous stairway from the highest to the lowest point in the building. The enclosure must be two-hour fire-resistive construction and shall only have those openings necessary for exiting from a normally occupied space. The enclosure shall be pressurized relative to the atmosphere, and the entrance vestibule and such ventilation equipment shall be provided with standby power.

Exit Passageways shall be as required by UBC 3312 and generally shall comply with 3305 for corridors. The construction shall be not less than one-hour fire-resistive with 3/4-hour protection of all openings, including doors. There shall be no openings other than required exits from normally occupied spaces.

UBC 3319 Exits - Group H Occupancies

"Every portion of a Group H Occupancy having a floor area of 200 square feet or more shall be served by at least two separate exits.

> **Exception:** Group H, Division 4 occupancies (Repair Garages) with floor area less than 1,000 square feet may have one exit.

Within Group H, Divisions 1, 2 and 3 Occupancies, all portions of any room shall be within 75 feet of an exit door. Exit doors from a room classified as Group H, Divisions 1, 2, and 3 Occupancies shall not be provided with a latch or lock unless it is panic hardware.

Doors leading to a corridor of fire resistive construction shall have a minimum of 3/4-hour fire-protection rating. . . shall be maintained self-closing . . . shall open in the direction of exit travel . . .

Within Group H, Division 7 and within fabrication areas of Group H, Division 6 Occupancies, the distance of travel to an exterior exit door, exit corridor, horizontal exit, exit passageway or an enclosed stairway <u>shall not exceed 100 feet.</u>"

Group H occupancies have special exiting requirements as defined in Section 3319. Travel distance from any point within a fab to an approved exitway or exterior door shall not exceed 100 feet. (The general requirement for Group B or other occupancies is 200 feet when the building is sprinklered.) Required number of exits from H occupancies is not based solely on occupant load, as is the case for most occupancies, but on room square footage. Under the 1985 UBC, any portion of an H occupancy over 200 square feet was required to have two exits. Under the 1988 UBC, H-3, H-4, H-5 and H-6 occupancies were erroneously required to have two (2) exits only where the room area exceeds 1,000 square feet. An errata was published to reduce this to 200 square feet, as originally intended. The 1991 UBC has returned to the more restrictive requirement of two exits for any room greater than 200 square feet in area. Check with your own corporate or insurance carrier safety policy, as many safety groups demand two exits in a room larger than 200 square feet, or <u>any</u> room of any size, where HPMs exist.

Exits must be no closer to each other than a distance of one-half the diagonal dimensions of the room. (This precludes two exits adjacent to each other, the intent being to maximize the likelihood that one exitway will be clear in an emergency). Another requirement unique to H Occupancies dictates a 3/4-hour fire rating on doors leading to exit corridors (UBC Section 911(b)1 and 3319), whereas these doors may be 20-minute rated in other occupancies.

The maximum allowable total travel distance from any location in a fab to the outside of a building, an enclosed stairway, a horizontal exit or an exit passageway may not exceed 300 feet, and then only when the last 100 feet are entirely within a one-hour fire-resistive exit corridor in compliance with UBC 3305, and the building is sprinklered (UBC 3303(d)). Therefore, if the travel distance from a point in the fab exceeds 300 feet, there must be an exitway other than an exit corridor at the end of the 300 feet travel distance, such as an exit passageway, horizontal exit area separation wall or enclosed stairway.

One exit path from a room in an "H" occupancy to an approved exit-way may pass through not more than one intervening room, where the adjoining room provides a direct, obvious and unobstructed means of travel to an exit corridor, exit enclosure or to the outside of the building (UBC 3303(e)).

> **Exception:** Rooms with cumulative occupant load of 10 or less may exit through more than one intervening room.

For illustrations of these requirements, refer to **Figures 2.7, 2.8, and 2.9.**

2.7   ALLOWABLE FABRICATION AREA (H-6)

Semiconductor fabrication areas (H-6 occupancy) are limited in size based upon the quantities of HPMs being used or stored. Article 51 of the Uniform Fire Code must be used in conjunction with the UBC when planning or evaluating permissible fabrication areas. Whereas Tables No. 9-A and 9-B in the UBC stipulate the maximum quantity of HPMs which may be used in a non-"H"-occupancy (control area), they do not stipulate the maximum amount of HPMs in a given "H" occupancy. UFC Tables 51.105-A and 51.105-B define the maximum quantity of HPMs allowed per fab area. The difference between the two tables is: Table 51.105-B stipulates absolute limitations, i.e., pounds, gallons or cubic feet, while 51.105-A stipulates density limits in lb./SF, gal./SF or CF/SF. Table 51.105-A

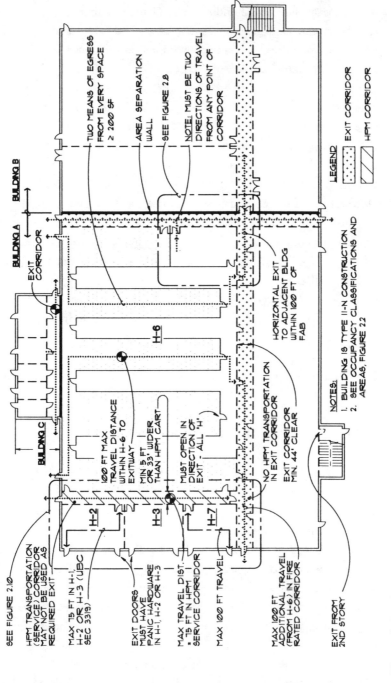

Figure 2.7 EXITING AND EXITWAYS are illustrated from each of the hazardous occupancies. Note requirements for two paths of egress and travel distance limitations.

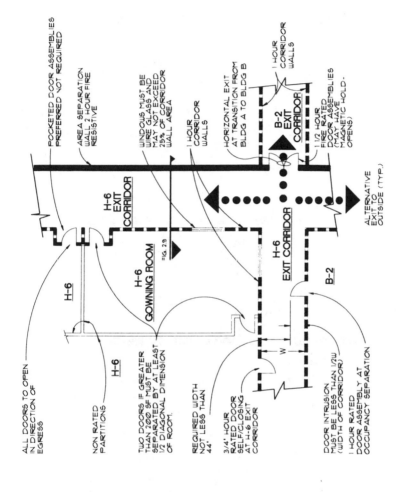

Figure 2.8 EXIT CORRIDOR REQUIREMENTS illustrating doors, glazing, width requirement and fire-resistive separations.

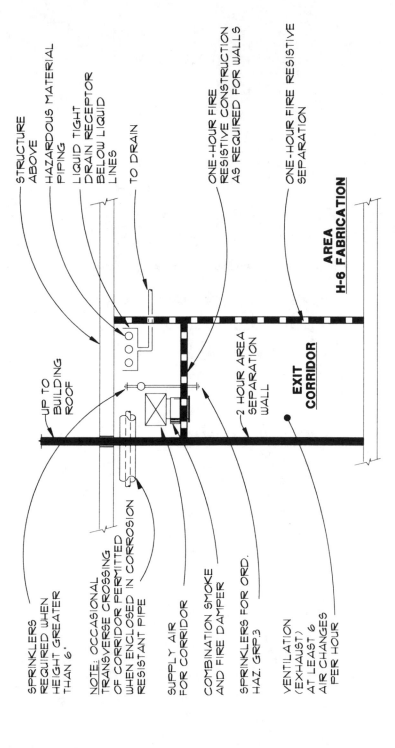

STRUCTURE
ABOVE

HAZARDOUS MATERIAL
PIPING

LIQUID TIGHT
DRAIN RECEPTOR
BELOW LIQUID
LINES

TO DRAIN

ONE-HOUR FIRE
RESISTIVE CONSTRUCTION
AS REQUIRED FOR WALLS

ONE-HOUR FIRE RESISTIVE
SEPARATION

**AREA
H-6 FABRICATION**

UP TO
BUILDING
ROOF

2 HOUR AREA
SEPARATION
WALL

**EXIT
CORRIDOR**

SPRINKLERS
REQUIRED WHEN
HEIGHT GREATER
THAN 6'

NOTE: OCCASIONAL
TRANSVERSE CROSSING
OF CORRIDOR PERMITTED
WHEN ENCLOSED IN CORROSION
RESISTANT PIPE

SUPPLY AIR
FOR CORRIDOR

COMBINATION SMOKE
AND FIRE DAMPER

SPRINKLERS FOR ORD.
HAZ. GRP.3

VENTILATION
(EXHAUST)
AT LEAST 6
AIR CHANGES
PER HOUR

Figure 2.9 EXIT CORRIDOR DETAIL.

further clarifies the disposition of HPMs in piping systems, stating such chemicals are not included in the computation. The intent of the code in Table 51.105-A is to allow for the creation of large fabrication areas that may use quantities of HPMs greater than those allowed in Table 51.105-B, the rationale being that the larger quantities are spread out over a large area, reducing the inherent risk of fuel contribution from one area to another.

Excessive amount of HPMs in a given existing facility (above that allowed by UFC 51.105-A or 51.105-B) may dictate the creation of two or more separated fab occupancies from a single existing fab. Options to creating multiple fabrication (H-6) occupancies would include:

1. Reduce the HPM inventory via "just-in-time" delivery techniques to comply with the Table 51.105 limits.
2. Create HPM storage rooms or areas of H-2, H-3 or H-7 classification, with proper occupancy separation from the H-6 to reduce the quantity of HPMs stored in the fabrication space.
3. Utilize bulk chemical distribution piping to deliver the HPMs from the storage rooms to the process tools. (The quantity of HPMs in piping is not included within the limitation on total quantity.)

2.8   MULTI-LEVEL FAB CONCEPT (H-6)

Fabrication areas must be separated from other fab areas, exit corridors and other parts of the building by not-less-than one-hour fire-resistive occupancy separations. The requirements for separation from other occupancies are outlined in UBC Table 5-B. The requirement for separation created the need for the concept of the "Multi-level Fab." By defining the mechanical support area of the fab (whether above or below, or both above and below, the actual working fab area) as a part of the fab environment, you may delete the need for an occupancy separation between the various levels of the fab and the support areas. This has a major impact on the mechanical systems because it eliminates the need for fire dampers in the ductwork penetrations of the main fab area. In order to qualify as a Multi-level fab, the interconnected level(s) must be used solely for equipment directly in support of the fab area (i.e., these cannot be occupied spaces). The language concerning the "multi-level" fab is found in the 1991 UBC Section 906; this exception was previously included in UBC 1706(a)8 (Shaft Enclosures).

UBC Sec. 1706(a) Classification of All Buildings by Types of Construction - Shaft Enclosures

"Openings through floors shall be enclosed in a shaft enclosure of fire-resistive construction having the time period set forth in Table No. 17-A for 'Shaft Enclosures,' except as permitted in Section 1706(c), (e) and (f). See Occupancy chapters for special provisions."

UBC 906 Requirements for Group H Occupancies - Shaft and Exit Enclosures

". . . In buildings with Group H, Division 6 Occupancies, a fabrication area may have mechanical, duct and piping penetrations which extend through not more than two floors within that fabrication area. The annular space around penetrations for cables, cable trays, tubing, piping, conduit or ducts shall be sealed at the floor level to restrict movement of air. The fabrication area, including the areas through which the ductwork and piping extend, shall be considered a single conditioned environment."

The code authorities anticipated a typical three-level design for a wafer fabrication facility. The top level will generally contain ducts and/or ventilation equipment which supply air to the working level of the fab. The lowest or basement level typically contains ducts, ventilation equipment, HPM piping and other services or equipment in support of the working level (*refer to* **Figure 2.6**). Fire dampers in the penetration of levels are deemed unnecessary as the ventilation air is generally part of the life safety system for the facility. Note, however, that seals are required at floor penetrations to prevent migration of air and smoke from one level to another (except, obviously, where the floor is used for return air to the lower level).

2.9   CORRIDORS

Applicable Code References

- Exit Corridors: (People Safety)
  - UBC Section 911(b)1
  - UBC Section 911(c)
  - UBC Section 3305

- Service Corridors: (Chemical Handling)
  - UBC Section 420 (Definition)
  - UBC Section 911(d)
  - UFC Section 51.109

2.9.1   Exit Corridors:

The requirements for exit corridors in hazardous occupancies are outlined in UBC 911(b)1, UBC 911(c), UBC 3305 and UFC 51.108, as discussed below. Exit corridors are intended for the safe use of people and are not intended to be used for transporting hazardous materials.

UBC 911(c)1   (Requirements for Group H Occupancy) Division 6 Occupancies - Exit Corridors
"Exit corridors shall comply with Section 3305 and shall be separated from fabrication areas as specified in Section 911(b).
1   Exit corridors shall not be used for transporting hazardous production materials except as provided in Section 911(f)2.

> **Exception:** In existing Group H, Division 6 Occupancies when there are alterations or modifications to existing fabrication areas, the building official may permit the transportation of hazardous production materials in exit corridors subject to the requirements of the Fire Code and as follows . . ." (See full text of code.)

UBC 911(f)2 (Division 6 Occupancies) Piping and Tubing - Installations in Exit Corridors and above Other Occupancies

"Hazardous production materials shall not be located within exit corridors or above areas not classified as Group H, Division 6 Occupancies, except as permitted by this subsection." (See full text of code.)

This provision of the UBC allowing a required exit corridor in an existing facility to be used for HPM transportation is an important issue in the design of a wafer fab retrofit. The codes recognize the difficulty which would be experienced by facility owners and designers if existing exit corridors could not be used for HPM transportation (as they frequently are in older wafer fab designs). Note, however, that the exit corridor must meet the requirements for an exit corridor and a service corridor.

- UFC 51.108(b) was specifically written to deal with the safety issues involved with transportation of HPMs in exit corridors in existing buildings.

- Exit corridors must be separated from other occupancies and the H-6 area by one-hour fire-rated construction. Door assemblies must be a minimum of 3/4-hour rated.

- Minimum exit width shall be 44 inches unless the occupant load is 10 or less, in which case the width may be 36" (UBC 3305).

- Air handling for the exit corridor must be separated from the fab air handling system. (*Refer to Chapter 4 of this book for further discussion.*)

- The exit corridor must not be used to return air from any other space.

- Exit doors must swing with egress (in the direction of travel into the corridor) and may not obstruct more than half of the clear opening of the corridor in any position and may not obstruct more than 7 inches of the required width when fully open (UBC 3305(d)).

- Door assemblies shall be tight-fitting smoke and have draft stops.

- Doors shall be self-closing or automatically-closing (magnetic hold open type) operated by smoke detector (per UBC 4306(b)).

- Glass in walls shall not exceed 25% of wall area and shall be 1/4" wire type.

- Duct penetrations shall be protected by dampers (UBC 4306(j)). (*Refer to Chapter 4 of this book for further discussion.*)

- Egress through the corridor shall be available in two directions except for a maximum dead-end of 20 feet.

Refer to **Figure 2.7** *for an illustration of these requirements.*

2.9.2 Service Corridors:

Service corridors are a special provision in Group H, Division 6 occupancies which are intended to provide for the safe transport of hazardous materials to the wafer fab from outside the building, or from an HPM storage room.

> **UBC 911(d) (Requirements for Group H Occupancy) Division 6 Occupancies - Service Corridors**
>
> "Service corridors shall be classified as Group H, Division 6 Occupancies. Service corridors shall be separated from exit corridors as required by Section 911(b)1.
>
> Service corridors shall be mechanically ventilated as required by Section 911(b)3 or at not less than six air changes per hour, whichever is greater.
>
> The maximum distance of travel from any point in a service corridor to an exterior exit door, horizontal exit, exit passageway, enclosed stairway or door into a fabrication area shall not exceed 75 feet. Dead-ends shall not exceed 4 feet in length. There shall be not less than two exits, and not more than one half of the required exits shall be into the fabrication area. Doors from service corridors shall swing in the direction of exit travel and shall be self-closing."

The subject of HPM transport in an exit corridor cannot be taken too lightly. A careful risk evaluation must be prepared with input from users, code officials, insurance carriers and the design team to determine the wisdom of mixing chemical delivery with people exiting the facility in an emergency.

(*There are many technical requirements related to HPM service corridors; these are discussed in Chapters 3, 4, 5, 6, and 7.*) In general, the requirements are:

■   Used for the transportation of HPMs in an "H" occupancy.

(Cont'd.)

- May not be used as required exit corridors, except in an existing H-6 where special allowances are made by building official (*see Section 2.9.1 above*).

- May not be crossed by required exit paths.

- Shall be classified as H-6 occupancy.

- Must be separated from exit corridors per 911(b)1 (one hour).

- Ventilation (exhaust) shall be one CFM/SF or six air changes/hour (min.).

- Maximum travel distance from any point to an exterior exit, horizontal exit (another "building"), exit passageway or enclosed stairway shall not exceed 75 feet (UBC 911(d)).

- Dead-ends shall not exceed four feet.

- Must have at least 2 exits and not more than half of the required exits shall be into the fabrication area.

- Minimum width shall be 5 ft. except that minimum must be 33" wider than the widest HPM cart or truck.

- All containers used to transport HPMs shall comply with UFC 51.108(b)2.

- Carts shall comply with 51.108(c) and shall not contain more than 55 gal. liquid or 7 gas cylinders (of up to 400 lbs.) or 500 lb. solids.

(Cont'd.)

- Maximum quantity of HPMs transported in service corridor at one time may be twice that stated in 51.108(c), in other words, up to two carts may be in the service corridor at a given time.

- Provide a local manual alarm station (spill alarm per 51.108(b)8).   This shall be fireman's telephone or pull station at not more than 150 ft. intervals and at each exit, with local signaling device and connection to Emergency Control Station.

- HPMs shall not be dispensed in the corridor.

- Pass-throughs (UFC 51.108(b)7) shall be allowed only in existing building exit corridors and shall be of one-hour construction with self-closing one-hour doors and shall be sprinklered.

- Sprinklers shall be designed for Ordinary Hazard Group 3 (as per H-6).

A clear identification of the need for, and location of, service corridors (where HPMs are transported) should be made by the designer early in the facility programming and layout process, in order to avoid problems with regulating agencies.   Again, service corridors are used for transporting HPMs.

Service corridors may not be used for required exiting.   Note, however, exit corridors in existing wafer fab buildings, which are to be altered or renovated, may be permitted by the building official to be used to transport HPMs, provided the exits comply with Sections 911(c) and UFC 51.108(b).

Figure 2.10 illustrates a service corridor which allows HPMs to be delivered to the fab without crossing the required exit corridors. Once inside the fab, HPMs may be distributed by the user at will, subject to the limitations on type of containers and carts outlined in the UFC.

Figure 3.2 illustrates a means used in a retrofit application to provide a service corridor adjacent to an existing exit corridor. The purpose of this retrofit was to eliminate chemical storage and pass-through cabinets which were previously installed in exit corridor walls.

2.10   CHEMICAL STORAGE

Due to the delivery schedule of hazardous production materials used in the fabrication process, it is often necessary to create storage or dispensing rooms within or adjacent to the fab area. The required fire-resistive separation rating is based on the size and specific classification of the storage room.

Storage rooms for liquid HPMs have additional requirements for liquid tight curbs or trenches with suitable drainage, to prevent spills from spreading beyond the limits of the storage area. Furthermore, in some storage facilities, spills must be conducted to a secondary containment area capable of holding the spill plus discharge from the fire sprinkling system until these fluids can be identified and neutralized.

When the size of a storage room dictates two exits, one exit must be directly to the exterior of the building.

The specific chemicals to be stored, their hazard rating and quantity, and the local code Authorities Having Jurisdiction will influence the design requirements. Experience has shown that a thorough dialogue with the building official is necessary, prior to going very far with the design of this type of facility.

GENERAL CORRIDOR REQUIREMENTS
- VENTILATE AT 1 CFM/SF OR 6 AIR CHANGES/HR
- SPRINKLERS PER ORDINARY HAZARD GROUP 3
- MINIMUM OF 2 EXITS NOT MORE THAN 1/2 REQUIRED
  EXITS TO FAB

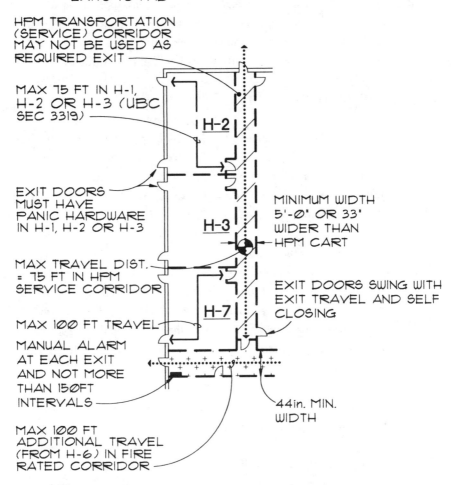

HPM TRANSPORTATION
(SERVICE) CORRIDOR
MAY NOT BE USED AS
REQUIRED EXIT

MAX 75 FT IN H-1,
H-2 OR H-3 (UBC
SEC 3319)

H-2

EXIT DOORS
MUST HAVE
PANIC HARDWARE
IN H-1, H-2 OR H-3

MINIMUM WIDTH
5'-0" OR 33"
WIDER THAN
HPM CART

H-3

MAX TRAVEL DIST.
= 75 FT IN HPM
SERVICE CORRIDOR

EXIT DOORS SWING WITH
EXIT TRAVEL AND SELF
CLOSING

H-7

MAX 100 FT TRAVEL

MANUAL ALARM
AT EACH EXIT
AND NOT MORE
THAN 150FT
INTERVALS

44in. MIN.
WIDTH

MAX 100 FT
ADDITIONAL TRAVEL
(FROM H-6) IN FIRE
RATED CORRIDOR

Figure 2.10 HPM STORAGE ROOMS AND HPM CORRIDOR.

A detailed discussion of the requirements for chemical storage rooms is beyond the scope of this book; however, some of the considerations are outlined in Chapter 3 - "Hazardous Material Storage and Handling." *Refer to Chapters 4, 5, 6 and 7 for more in-depth discussions of mechanical and electrical requirements in HPM storage facilities.*

SUMMARY

ARCHITECTURAL ISSUES (2.0)

♦   "Architectural issues" - deal with the fundamental organization of a hazardous occupancy and its context with the rest of a facility.

RENOVATION (2.2)

♦   Many industrial facilities were designed as B-2 occupancies, prior to the advent of UBC occupancy classifications H-3, H-6 and H-7.

♦   UBC B-2 does not address the subject of allowable floor area with respect to chemical usage; prior to 1985, the facilities were allowed to be of unlimited area, provided they were sprinklered and had side yards. Provisions of UBC Chapter 9 (Hazardous Occupancies) and UFC Article 51 limit the allowable areas of these occupancies with respect to types and quantities of chemicals used.

♦   The extent of a renovation, addition, or change in function, which may be tolerated before the entire building needs to be brought into compliance, is a matter of the interpretation of the local building official and/or fire marshal. The knowledgeable facility owner can influence this by negotiation of alternative strategies.

OCCUPANCY CLASSIFICATION (2.3)

♦   The first decision which must be made during the design process concerns the correct occupancy classification(s) for the proposed facility. A summary of the quantity, location and type of hazardous materials located within a given "control area" is required in order to make the correct occupancy classifications.

♦   In the event a review of the HPM inventory indicates quantities above the "exempt" amounts set in UBC Tables 9-A or 9-B for a control area, an "H" occupancy classification is mandated.

◆ UFC Tables 51.105-A and 51.105-B establish the maximum amount of HPMs in a single fabrication area (H-6), both in absolute quantity and maximum density.

◆ Mixed occupancies often exist in a given facility or building. In a building where the area of a different occupancy is less than 10 percent of the area of the primary occupancy, a mixed occupancy is not considered to occur; this is considered a "minor accessory use" condition.

◆ When a different occupancy exceeds 10 percent of the area of the primary occupancy, UBC considers the building a mixed occupancy and requires the summation of the ratio of actual area to allowable area for each occupancy to add up to one or less.

◆ When more than one occupancy occurs within a building, fire-resistive occupancy separations must be constructed between them (UBC Table 5-B and Sections 503(b), (c), and (d)).

◆ Note that an H-1 occupancy may <u>not</u> be mixed with any other occupancy.

◆ <u>H-7 Occupancies</u>

   ▪ The 1988 Uniform Building Code identified the category of "H-7" occupancy. H-7 is a classification for facilities which store and use large quantities (above the "exempt" amounts) of materials which are considered "health hazards."

   ▪ Research laboratories, pilot production facilities, plating shops and the like may be appropriately classified as H-7 occupancies.

ALLOWABLE AREA AND SEPARATIONS (2.4)

◆   Facilities which were originally constructed as B-2 occupancy were allowed to have unlimited area, when sprinklered and surrounded by side yards. As such facilities are converted to Group "H" occupancy, severe restrictions on the allowable area may be imposed by the UBC.

◆   The maximum allowable area must be evaluated based upon the correct occupancy and construction type in conjunction with UBC Table 5-C and UBC Sections 505 and 506.

   ▪   The areas specified in Table No. 5-C and Section 505(b) may be tripled in one-story buildings and doubled in buildings of more than one story if the building is provided with an approved automatic sprinkler system throughout.

   ▪   The basic allowable area may be doubled for multiple story buildings.

   ▪   The basic allowable area may be increased for side yards (accessible to the fire department) where they exceed 20 feet.

   ▪   The limitations in the UBC and UFC on allowable quantities of HPMs in a given occupancy must be evaluated in concert with the area test.

◆   Many wafer fabrication occupancies have basements used for the support of fabrication operations. If the lowest level is <u>not</u> considered a first-story (*see definitions in Section 403 and 420*), then it does not enter into the calculation of allowable area, regardless of its occupancy.

◆   Separation of an existing structure into <u>separate buildings</u> (if required to comply with the maximum allowable area limitations) may be accomplished by the construction of <u>area separation walls</u> (Section 505(f)).

LOCATION (2.5)

◆ The location of various occupancies, with respect to the story and height within a building, the proximity to a property line, etc., is generally governed by UBC Chapter 5.

■ UBC Table 5-A stipulates the fire ratings of exterior walls and openings in walls.

■ UBC Table 5-D stipulates the maximum height of a building containing each type of occupancy.

◆ UBC Table 9-E stipulates the quantities of explosives, oxidizers, reactives, unstable materials and pyrophoric gases, which, when exceeded, must be stored in a detached building.

◆ UBC 51.105(b) states that occupied levels of fabrication areas shall be located at, or above, the first story.

EXITING (2.6)

◆ General requirements for exiting:

■ **UBC 3301(b) Exits - Definitions**
"EXIT is a continuous and unobstructed means of egress to a public way and shall include . . ."

■ **UBC 3303(e) Exits Required - Exits Through Adjoining Rooms**
"Rooms may have one exit through an adjoining or intervening room . . ."
Exceptions:
2 Rooms with a cumulative occupant load of 10 or less may exit through more than one intervening room."

◆   Exit corridors shall be arranged to allow travel from any point in two directions to an exit, except for dead-ends not exceeding 20 feet in length.

◆   Horizontal Exits shall be as specified in UBC 3308. Horizontal exit is an exit from one building into another building or through or around a wall constructed as required for a two-hour occupancy separation.

◆   Required exit doors must be no closer to each other than a distance of one-half the diagonal dimensions of the room.

◆   UBC Section 3319 gives specific requirements for H occupancies.

▪   **UBC 3319 Exits - Group H Occupancies**
"Every portion of a Group H Occupancy having a floor area of 200 square feet or more shall be served by at least two separate exits.

▪   "....Within Group H, Division 7 and within fabrication areas of Group H, Division 6 Occupancies, the distance of travel to an exterior exit door, exit corridor, horizontal exit, exit passageway or an enclosed stairway shall not exceed 100 feet....."

▪   "....The maximum allowable total travel distance from any location in a fab (H-6 or H-7)to the outside of a building, an enclosed stairway, a horizontal exit or an exit passageway may not exceed 200 feet, and then only when the last 100 feet are within an exit corridor in compliance with UBC 3305, and the building is sprinklered (UBC 3303(d))...."

ALLOWABLE FABRICATION AREA (H-6) (2.7)

◆   Fabrication areas (H-6 occupancy) are limited in size based upon the quantities of HPMs being used or stored. UFC Tables 51.105-A and 51.105-B define the maximum quantity of HPMs allowed per fab area.

- Table 51.105-B stipulates absolute limitations, i.e., pounds, gallons or cubic feet.
- Table 51.105-A stipulates density limits in lb./SF, gal./SF or CF/SF.

♦ Excessive amount of HPMs in a given existing facility (above that allowed by UFC Tables 51.105-A or 51.105-B) may dictate the creation of two or more separated fab occupancies from a single existing fab. Options to creating multiple fabrication (H-6) occupancies would include:

- Reduce the HPM inventory via "just-in-time" delivery.
- Create HPM storage rooms or areas of H-2, H-3 or H-7 classification, with proper occupancy separation from the H-6.

MULTI-LEVEL FAB CONCEPT (H-6) (2.8)

♦ Fabrication areas must be separated from other fab areas, exit corridors and other parts of the building by not-less-than one-hour fire-resistive occupancy separations.

♦ By defining the mechanical support area of the fab as a part of the fab environment, you may delete the need for an occupancy separation between the various <u>unoccupied</u> levels of the fab and the support areas.

♦ UBC 906 eliminates the need for fire dampers in the ductwork penetrations of the various levels of such a multi-level fabrication facility. All levels are considered part of the same fire zone.

EXIT CORRIDORS: (2.9.1)

♦ Exit corridors are designed to provide for safe egress of people and are governed by UBC 911(b)1, 911(c)1, UBC 3305 and UFC 51.108.

◆   **UBC 911(c)1**

"Exit corridors shall comply with Section 3305 and shall be separated from fabrication areas.

1   Exit corridors shall not be used for transporting hazardous production materials except as provided in Section 911(f)2.

> **Exception:** In existing Group H, Division 6 Occupancies when there are alterations or modifications to existing fabrication areas, the building official may permit the transportation of hazardous production materials in exit corridors subject to the requirements of the Fire Code and as follows . . ." (See full text of code.)

◆   UFC 51.108(b) was specifically written to deal with the safety issues involved with transportation of HPMs in exit corridors in existing buildings.

◆   There are many special requirements for exit corridors in an H occupancy:

- Air handling for the exit corridor must be separated from the fab air handling system.
- Exit doors must swing with egress.
- Door assemblies shall be tight-fitting smoke and have draft stops.
- Doors shall be self-closing or automatically-closing and shall be 3/4 hour rated.
- Duct penetrations shall be protected by fire and smoke dampers.
- Egress through the corridor shall be available in two directions.

SERVICE CORRIDORS: (2.9.2)

◆   Service corridors are intended for the transportation of HPMs with H-6 occupancy and are governed by UBC 420, 911(d), UFC 51.109:

◆   **UBC 911(d)**

"...Service corridors shall be classified as Group H, Division 6 Occupancies. Service corridors shall be separated from exit corridors as required by Section 911(b)1.

The maximum distance of travel from any point in a service corridor to an exterior exit door, horizontal exit, exit passageway, enclosed stairway or door into a fabrication area shall not exceed 75 feet..".

◆   There are many technical requirements related to HPM service corridors:

- Used for the transportation of HPMs in an "H" occupancy.

- May not be used as required exit corridors, except in an existing H-6.

- May not be crossed by required exit paths.

- Must be separated from exit corridors per 911(b)1 (one hour).

- Ventilation (exhaust) shall be one CFM/SF or six air changes/hour (min.).

- Must have at least 2 exits.

- All containers used to transport HPMs shall comply with UFC 51.108(b)2.

- Provide a local manual alarm station (spill alarm per 51.108(b)8).

- HPMs shall not be dispensed in the corridor.

- Pass-throughs shall be allowed <u>only</u> in existing building exit corridors (UFC 51.108(b)7) .

◆   <u>Once inside the fab, HPMs may be distributed by the user at will, subject to the limitations on type of containers and carts outlined in the UFC.</u>

3

Hazardous Materials Storage and Handling

3.1 INTRODUCTION

The subject of storage and handling of hazardous materials is a complex one. Hazardous Production Materials (HPMs) are strictly regulated by the model codes and local jurisdictions. The use of these materials represents potential danger to life and property; therefore, the liability of users of HPMs is significant. Protection of life and property is the primary goal of the codes and of building owners. This chapter will deal with the requirements for storage and handling facilities as provided in the UFC (Articles 51, 79 and 80) and the UBC (Chapter 9). For the purpose of this chapter of the book, it will be assumed that we are dealing with an occupancy with quantities of hazardous materials in excess of the "exempt" amounts as defined in UBC Tables 9-A and 9-B.

As outlined in Chapter 1, there is a proposal to create a special UFC Standard 80-1 to deal with the special requirements and hazards (pyrophoric/unstable) associated with silane and gas mixtures which contain more than 2 percent silane. The proposed standard addresses specific requirements for fire protection, restricted flow orifices, high velocity ventilation and separation of cylinders. This text will not address the specific requirements of Standard 80-1, as this is not yet a binding regulation. We would encourage all those involved with the design and/or operation of silane facilities to review the proposed text of 80-1 with regard to implications it may have on your project and/or facility. If adopted by the Uniform Fire Code committee, the new language could be published in the 1994 UFC.

A complete discussion of all the design and operating requirements of the various code sections related to HPM storage and handling goes beyond the scope of this book. As each facility and project is unique, the methods of providing code compliance will necessarily be unique to each project.

Highlights of the various codes, specifically pertaining to semiconductor manufacturing facilities will be provided. *In addition, more specific technical discussion about design features related to HVAC, Fire Suppression, Power and Alarm/Monitoring is provided in Chapters 4, 5, 6 and 7, respectively.*

3.2　HAZARDOUS MATERIALS

3.2.1　Definitions: The following definitions and/or brief code excerpts are provided to illustrate the overlapping provisions of the UFC and UBC model codes. These should be used (see the full text of the code) when analyzing any situation involving HPMs.

> "HAZARDOUS PRODUCTION MATERIAL (HPM) is a solid, liquid or gas that has a degree-of-hazard rating in health, flammability or reactivity of Class 3 or 4 as ranked by UFC standard No. 79-3 and which is used directly in research, laboratory or production processes which have as their end-product materials which are not hazardous." **(UFC 9.110)**

> "STORAGE is the keeping, retention or leaving of flammables or combustible liquids in closed containers, tanks or similar vessels." (Although UFC 79 relates specifically to flammables, it would be safe to assume the same definition can also apply to other HPMs.) **(UFC 79.102(b))**

> "USE (Material) is the placing in action or making available for service by opening or connecting anything utilized for confinement of material whether a solid, liquid or gas." **(UFC 9.123)**

> "DISPENSING is the pouring or transferring of any material from a container, tank or similar vessel whereby vapors, dusts, fumes, mists or gases may be liberated to the atmosphere." **(UFC 9.106)**

> "DISPENSING AND MIXING. Liquids shall not be dispensed or mixed in liquid storage rooms unless such rooms comply with the electrical, heating and ventilation requirements in Division VIII (of Article 79). Liquid storage rooms in which liquids are used, dispensed or mixed shall be classified as Group H, Division 2 Occupancies." **(UFC 79.203(f))**

"WORKSTATION is a defined space or independent principal piece of equipment using HPMs within a fabrication area where a specific function, a laboratory procedure or a research activity occurs. Approved cabinets serving the workstation are included as a part of the workstation. A workstation may contain ventilation equipment, fire protection devices, sensors for gas and other hazards, electrical devices and other processing and scientific equipment." **(UFC 51.102(b))**

STORAGE CABINET requires that the storage of HPM liquids, gases and solids in the fabrication area is to be within fully-enclosed storage cabinets or within a workstation. Construction and operation requirements, gas detection, controls and sprinkler requirements are called out in this section. Requirements for storage cabinets for flammable and combustible liquids are called out in UFC 79.202. **(UFC 51.107)**

"HPM STORAGE ROOM is a room used for the storage or dispensing of HPMs and which is classified as a Group H, Division 2, 3 or 7 Occupancy." (Construction requirements including the size, location and separation wall requirements for HPM storage rooms are given in UBC Section 911(e)). **(UFC 9.110)**

"INSIDE HPM STORAGE ROOM is an HPM storage room which is totally enclosed within a building and having no exterior walls." (Again, construction requirements including the size, location and separation wall requirements for inside HPM storage rooms are given in UBC Section 911(e)). **(UFC 9.111)**

"EMPTY CONTAINERS. The storage of empty tanks and containers previously used for storage of flammable or combustible liquids, unless free from explosive vapors, shall be as specified for the storage of flammable liquids. Tanks and containers when emptied shall have the covers or plugs immediately replaced in openings."

Although this excerpt pertains strictly to flammable and combustible liquids, we recommend that any HPM empty containers be treated the same as a full container, unless specifically exempted by the codes. **(UFC 79.201(c))** (*Refer to* **Figure 3.1** *for types of containers typically encountered.*)

Figure 3.1 VARIOUS CONTAINERS USED FOR HAZARDOUS CHEMICALS. (Photo courtesy of Olin Hunt Co.)

COMBUSTIBLE LIQUID is a liquid having a flash point at or above 100°F....

> Class II: 100° to 140°F;
> Class III-A: 140° to 200°F;
> Class III-B: at or above 200°F;
> (all flash points). **(UFC 9.105)**

FLAMMABLE LIQUID is a liquid having a flash point below 100°F and having a vapor pressure not exceeding 40 psia at 100°F.

> Class I-A: flash point below 73°F and boiling point below 100°F;
> Class I-B: flash point below 73°F and boiling point at or above 100°F;
> Class I-C: flash point at or above 73°F and below 100°F. **(UFC 9.108)**

3.2.2 Classification of HPMs:

For guidance on the classification of chemicals by hazard category, refer to Appendix VI-A of the UFC. This appendix gives descriptions of each classification and examples of materials falling into each class. For additional assistance, the supplier of each material should be consulted. Manufacturers are required to publish "Material Safety Data Sheets" (MSDS) which give complete technical information on their products.

Article 80 of the UFC provides general requirements for all types of hazardous materials regardless of the occupancy or use of a facility. In the event that specific applications for process materials are not fully defined in other applicable "Group H" codes, Article 80 will govern.

Article 80 subdivides hazardous materials classifications into two broad categories: physical hazards and health hazards (UFC 80.202). Definitions of these categories are provided in UFC Article 9; further discussion and examples are provided in Appendix VI-A. Physical and health hazards possess any of the following properties or characteristics:

Physical Hazards

1.   Explosive and blasting agents, regulated under Article 77.
2.   Compressed gases, regulated under Article 80 and Article 74.
3.   Flammable and combustible liquids regulated under Article 79.
4.   Flammable solids.
5.   Organic peroxides.
6.   Oxidizers.
7.   Pyrophoric materials.
8.   Unstable (reactive) materials.
9.   Water-reactive materials.
10.   Cryogenic fluids, regulated under Article 80 and Article 75.

<u>Health Hazards</u>

1.   Highly toxic or toxic materials, including highly toxic or toxic compressed gases.
2.   Radioactive materials.
3.   Corrosives.
4.   Carcinogens, irritants, sensitizers, and other health hazards.

Under UFC Section 80.101(b), carcinogens, irritants, and sensitizers do <u>not</u> include commonly used building materials and consumer products which are not otherwise regulated by the UFC.

Materials with a primary classification as a physical hazard may also present a health hazard, and materials with a primary classification as a health hazard may also present a physical hazard. Compressed diborane and phosphine are examples of materials that are both a physical hazard and health hazard. When a given material falls into multiple hazard categories, all of the hazards must be addressed by the facility design (see UFC Section 80.101(c)).

3.3   MAXIMUM ALLOWABLE QUANTITIES OF HPMS

3.3.1   General:

It is interesting to note that, with few exceptions, the maximum quantities of hazardous materials permitted within H-1, H-2, H-3 or H-7 occupancies are not limited. The exceptions are the area limitations on occupancies provided in the UBC (Table 5-C) and certain gross limitations on the size of hazardous material stockpiles in UFC Article 80. Group H, Division 6 occupancies are limited in the maximum amount of hazardous materials within a "single fabrication area" as discussed below.

In the case of certain chemicals which can undergo rapid thermal reactions or decomposition (such as pyrophorics, explosives or other detonatables, water reactives, unstable reactives, oxidifiers and organic peroxides) consult UBC Table 9-E for the maximum storage limit in the building (*provided for your convenience as* **Table 3.1.**). If you exceed these values, a storage facility detached from the main building is required.

OSHA REGULATIONS: The Occupational Safety and Health Administration (OSHA) added a requirement in May of 1992 for those facilities that maintain toxic or reactive chemicals in excess of certain quantities. The new standard, under 29 CFR Part 1910.119, requires establishment of a process safety management (PSM) program. Owners are required to analyze their processes, identify potential trouble spots and develop measures to prevent chemical releases. The intent of this regulation is directed, not at forcing users to reduce chemical quantities through increased regulations, but instead at encouraging users to take a "hard look" at their processes and make them safer.

Typical Hazardous Production Materials (HPMs) used in cleanroom facilities that may be found on the OSHA list (and their respective limits) are: ammonia, 10,000 pounds; arsine, 100 pounds; chlorine, 1,500 pounds, and hydrogen fluoride, 1,000 pounds. These values are considerably in excess of the "exempt amounts" established by national building codes, such as the Uniform Building Code or the BOCA National Building Code, for a hazardous occupancy. If the HPM amounts within a facility can be held to under the OSHA amounts, establishment of a government-regulated PSM program will not be required.

TABLE NO. 9-E—REQUIRED DETACHED STORAGE

DETACHED STORAGE IS REQUIRED WHEN THE QUANTITY OF MATERIAL EXCEEDS THAT LISTED			
Material		**Solids and Liquids (Tons)[1,2]**	**Gases (Cubic Feet)[1,2]**
1. Explosives, blasting agents, black powder, fireworks, detonatable organic peroxides, 2. Class 4 oxidizers, 3. Class 4 or Class 3 detonatable unstable (reactives)		Over exempt amounts	Over exempt amounts
4. Oxidizers, liquids and solids	Class 3 Class 2	1,200 2,000	— —
5. Organic peroxides	Class I Class II Class III	Over exempt amounts 25 50	— — —
6. Unstable (reactives)	Class 4 Class 3 Class 2	1/1,000 1 25	20 2,000 10,000
7. Water reactives	Class 3 Class 2	1 25	2,000 10,000
8. Pyrophoric gases		—	2,000

[1]Distance to other buildings or property lines shall be as specified in Table No. 9-D based on TNT equivalence of the material.
[2]Over exempt amounts mean over the quantities listed in Table No. 9-A.

Table 3.1   UBC REQUIREMENTS FOR SEPARATED STORAGE (UBC Table 9-E)

3.3.2   H-6 Occupancies (General Storage/Use):

There are two methods of calculating the maximum allowable quantity of each category of HPMs in a single H-6 fabrication occupancy:

Absolute Quantity: UFC Table 51.105-B stipulates the maximum permissible quantities (in gallons, pounds, or cubic feet) of various categories of HPMs which may be contained within a given H-6 occupancy, regardless of the size of the facility.   Unlike UBC Table 9-A where the exempt quantities may be increased for fire sprinklers and storage in safety cabinets,   Table 51.105-B does not allow increases.   Note also that highly toxic materials and toxic or highly toxic gases are to be included in the aggregate maximums for flammable materials.

Provided for your convenience are UFC Tables 51.105-A and 51.105-B as our **Table 3.2.**

TABLE NO. 51.105-A—AVERAGE DENSITY OF HPM IN A SINGLE GROUP H, DIVISION 6 OCCUPANCY[1, 2]

STATE	UNITS	FLAMMABLE	OXIDIZER	CORROSIVE
Solid	lb./sq.ft.	0.001	0.003	0.003
Liquid	gal./sq.ft.	0.04[3]	0.03	0.08
Gas	c.f./sq.ft.	2	1.250	3.000

[1]Hazardous production materials within piping shall not be included in the calculated quantities.

[2]The maximum permitted quantities of gases which are flammable, toxic or highly toxic shall not exceed the quantity specified by Table No. 51.105-B.

[3]The average densities of flammable and combustible liquids shall not exceed the following:

Class (I-A) + (I-B) + (I-C) (Combination flammable liquids) . . . . = .025
    however, Class I-A shall not exceed . . . . = .0025

Class II . . . . = .01

Class III-A . . . . = .02

TABLE NO. 51.105-B—PERMITTED QUANTITIES OF HPM IN A SINGLE FABRICATION AREA

MATERIAL	MAXIMUM QUANTITY
Flammable liquids	
Class I-A	90 gal.
Class I-B	180 gal.
Class I-C	270 gal.
Combination flammable liquids	360 gal.
Combustible liquids	
Class II	360 gal.
Class III-A	750 gal.
Flammable gases	9,000 cu. ft. at one atmosphere of pressure at 70°F.
Liquefied flammable gases	180 gal.
Flammable solids	1,500 lbs.
Corrosive liquids	165 gal.
Oxidizing material—gases	18,000 cu.ft.
Oxidizing material—liquids	150 gal.
Oxidizing material—solids	1,500 lbs.
Organic peroxides	30 lbs.
Highly toxic material and toxic or highly toxic gas	Included in the aggregate for flammables as noted above regardless of flammability characteristics

Table 3.2   MAXIMUM PERMITTED HPMs IN H-6 (1991 Uniform Fire Code Tables No. 51.105-A and 51.105-B).

Density Limit: UFC Table 51.105-A stipulates limits for the density in lb./SF, gal./SF or CF/SF of HPM solids, liquids and gases which may be stored, used or dispensed in a single H-6 fabrication area. The quantity of HPM within closed piping systems in not included in the calculations. All other HPMs are included in the maximum allowance calculations and might include:

- in-use at workstation.
- unsegregated storage.
- segregated storage.

The density limitation does not preclude any fabrication area, regardless of size from having the maximum quantity of HPM allowed by Table 51.105-A, thus small H-6 occupancies are not penalized by the density limitation.

3.3.3   H-6 Occupancies, In-Use Within Workstations:

The quantities of chemicals in-use within a workstation are included in the total amount allowed within the fab. The maximum quantities of HPMs in use at any one workstation per Table 51.106-A in UFC 51.106(b) shall not exceed the following:

Flammables + Toxics (combined)	gases	3 cylinders
	liquids	15 gallons
	solids	5 pounds
Corrosives	gases	3 cylinders
	liquids	25 gallons*
	solids	20 pounds

Oxidizers	gases	3 cylinders
	liquids	12 gallons*
	solids	20 pounds

*   (An equal amount of nonflammable HPM liquid in reservoirs of filtering systems of connected materials in use is permitted.)

UFC Section 51.106 discusses construction requirements for workstations. Storage cabinets for HPM gases and flammable liquids shall conform to the requirements outlined in UFC 51.107(b), UFC 80.303(a)6B and UFC 79.202(c).

3.3.4   Piped Systems:

Quantities of HPMs in utility, process or waste pipes are not included in quantity limitations. Intermediate reservoirs and pumping stations, however, are not defined as "pipes" even though they comprise part of a completely closed system. Quantities in reservoirs must, therefore, be counted against the total allowed in storage.

3.3.5   H-6 Occupancies, Unsegregated Storage Within Fab:

Unsegregated storage of HPMs (not separated by rated construction) within the fab area is allowed under certain quantities and storage conditions. In general, HPMs may be stored in the fab if quantities are within the limits of UFC Tables 51.105-A or -B. Allowable quantities of flammable and combustible liquids are governed by UFC Article 79.

UFC Section 51.107 regulates HPMs within the fab stored outside of workstations. Similar to Section 51.106 which regulates requirements for workstations, there are construction requirements for the cabinets used to store HPMs. However, unlike the treatment for workstations, Section 51.107 does not specify limits on the quantities of HPMs allowed in each cabinet.

Storage of HPMs in excess of the quantities in UBC Tables 9-A and 9-B shall be in a room complying with the requirements of the Building Code for an inside liquid room or in an HPM storage room (*see requirements for segregated storage rooms following*).

The maximum exempt unsegregated quantities of flammable and combustible liquids which may be stored in a control area are listed in UFC Table No. 79.202-A. (*Control areas are further discussed in Section 3.5.*) Quantities listed allow for a 100 percent increase for fully sprinklered buildings.

Class I-A	60 gallons
Class I-B	120 gallons
Class I-C	180 gallons
Class II	240 gallons
Class III-A	660 gallons

The total for a combination of Class I and II liquids and for any combination of Class I liquids may not exceed 240 gallons, subject to the limitations of each individual class. Quantities of Class III-B combustible liquids are not limited, when stored in buildings equipped with fully automatic sprinklers. Quantities exceeding these limits must be within liquid storage rooms or liquid storage warehouses.

UFC Article 79.202(c)2 also limits the amounts per storage cabinet. Class I or Class II liquids may not exceed 60 gallons, and the total quantities of all liquids shall not exceed 120 gallons. UFC 79.202(c)4(ii) allows no more than three cabinets in a room, unless each group of three is separated by a minimum of 100 feet.

3.3.6   Segregated Storage (HPM Storage Rooms) - within the H-6:

UBC 911(e)1 states that stored quantities of HPMs exceeding the maximum amounts listed in UBC Tables 9-A or 9-B must be in a room complying with UBC requirements for an inside liquid storage room or an HPM storage room. Flammable or combustible liquids which exceed UFC Table 79.202-A require the creation of separate HPM storage rooms or inside storage rooms. Where quantities of materials constituting a health hazard (such as corrosives, i.e., acids or bases) exceed the quantities listed in UBC Table 9-B, a separate H-7 storage room must be created. The requirements for storage of HPMs other than flammable/combustible liquids are detailed in UFC Article 80, with general requirements for storage in Division III and general requirements for dispensing/use and handling in Division IV.

These storage rooms provide separation, protection, size limitation and location such that the stored HPM is not an additional hazard to the fab. Because there is a separation, the total quantity of various HPMs within the fab may exceed the limits given in UBC Tables 9-A, 9-B and UFC 79.202-A; however, the total amounts (including storage room amounts) must still be within the amounts allowed for the entire fab per UFC Tables 51.105A and 51.105B.

The allowable quantities within each defined HPM storage room are limited to the quantities specified in the appropriate UFC Articles, in particular, Articles 79 and 80.

In addition, UFC 51.110(b)4 restricts the quantities within HPM storage rooms as follows:

- Highly toxic liquids shall be counted as flammable liquids.
- Corrosive liquids shall be counted as Class III flammable liquids.
- Highly toxic solids shall be counted as flammable solids.
- The quantities of HPM gases shall not exceed the following:

Oxidizers	20,000 cubic feet
Corrosives	30,000 cubic feet
Flammables	15,000 cubic feet
Toxics + High toxics	Counted as flammable gases

(Note: A separate chemical storage occupancy (other than H-6 occupancy) is not considered to be part of the fab by virtue of the occupancy separation wall; it is, therefore, not limited by any provisions within Article 51.)

The requirements for the construction and operation of HPM storage rooms are complex and outlined in detail in the various UBC and UFC provisions. *(An introduction to these requirements is provided in Section 3.6 of this book.)*

3.4   TRANSPORTATION OF HPMs

3.4.1   Manual Transportation:

Manual transportation of HPMs <u>within a single fab area</u> is not specifically limited other than requiring the use of approved chemical containers.

Manual transportation of HPMs between fabs, or between segregated storage rooms, however, requires the creation of a service corridor. Many existing wafer fabs do not have service corridors, as the fabs were designed as one entity with no divisions, control areas or segregated storage areas. Generally, service corridors impose severe limitations on the fab layout because <u>they cannot serve as exit corridors</u>, cannot be crossed with any exit paths, and require closely spaced access to exits.

UFC 51.108(a) stipulates that HPMs are <u>not permitted in exit corridors of new buildings</u>. In this case, exit corridors are those <u>required</u> exitways, and should not be confused with service aisles or other walkways or circulation space within the fab occupancy.

In <u>existing buildings</u> where the exit corridors must be used for transportation of HPMs (which is often the case in older fabs), this practice may be continued, provided the owner complies with the requirements of UFC 51.108(b) and UBC 911(c).   Manual transportation within exit corridors must be in approved chemical carts.

Figure 3.2 illustrates a retrofit of an existing facility which had HPM liquid pass-through cabinets located in an exit corridor. As a means of <u>mitigating</u> this hazard, the design team negotiated a compromise compliance strategy. The exit corridor continued to be utilized for the transport of HPMs (under UBC 911(c) and UFC 51.108(b)) and the pass-through cabinets were relocated to a fire-separated HPM "service" corridor adjacent to the fabrication area. Such compromise strategies, while not in full compliance with the <u>letter</u> of the new code provisions, help to satisfy the intent, thus reducing risk to the occupants.

(A discussion of the requirements associated with service corridors is provided in Chapter 2 of this book.)

3.4.2   Piping Systems for HPMs:

Enclosed piping systems used to transport HPMs to or through the fab must meet the strict requirements of multiple UBC and UFC codes. *(Although a thorough discussion of HPM piping requirements is beyond the scope of this book, Section 3.8 provides a compendium of the UBC and UFC code requirements for HPM piping systems.)* Note that, where the piping passes over exit corridors, additional requirements (UBC 911(f), UFC 51.105(e)2 and 51.108(d)) come into effect.

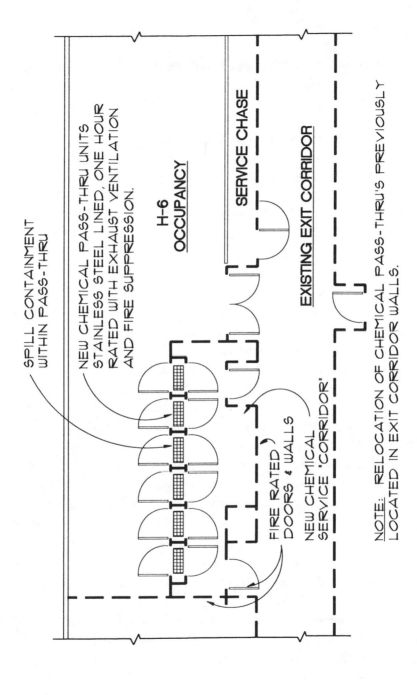

SPILL CONTAINMENT WITHIN PASS-THRU

NEW CHEMICAL PASS-THRU UNITS STAINLESS STEEL LINED, ONE HOUR RATED WITH EXHAUST VENTILATION AND FIRE SUPPRESSION.

H-6 OCCUPANCY

SERVICE CHASE

EXISTING EXIT CORRIDOR

FIRE RATED DOORS & WALLS

NEW CHEMICAL SERVICE 'CORRIDOR'

NOTE: RELOCATION OF CHEMICAL PASS-THRU'S PREVIOUSLY LOCATED IN EXIT CORRIDOR WALLS.

Figure 3.2 RETROFIT OF HPM PASS-THROUGHS From Existing Exit Corridor to New Service Corridor.

3.5   CONTROL AREAS

3.5.1   Areas within a facility which use or store less than the "exempt" amounts of HPMs may be considered "control areas" and must be constructed with not less than one-hour partitions.

> **UFC 9.105** defines the concept of a control area:
> "CONTROL AREA is a space within a building where the exempt amounts (of HPMs) may be stored, dispensed, used or handled."

The UBC defines a control area in Section 404 and in Footnote 1 of Tables 9-A and 9-B as follows:

> "CONTROL AREA is a space bounded by not less than a one-hour fire-resistive occupancy separation within which the exempted amounts of hazardous materials may be stored, dispensed, handled or used . . . The number of control areas in buildings with other uses shall not exceed four."

3.5.2   The "control area" is a space within an occupancy which does <u>not</u> have to be considered a specific hazardous occupancy such as H-1, H-2, H-3, H-6 or H-7. The codes allow an owner to have exempt quantities of hazardous materials in a room with a one-hour fire separation, but without the other facilities required for a hazardous storage occupancy, such as ventilation, alarms, emergency power, etc. There may be up to four individual control areas within a building (other than a retail store); however, the aggregate quantity of HPMs in a fab area (including control rooms) may not exceed the maximum amounts in UFC Table 51.105-A or 51.105-B.

If the quantities of hazardous materials exceed the exempt amounts given for various (flammable or combustible) material types in UFC Article 79 (as discussed previously) or Division III and IV of Article 80 (Tables 80.303-A, 80.305-A, 80.306-A, 80.307-A, 80.308-A, 80.309-A, 80.310-A, 80.312-A, 80.313-A, 80.314-A, 80.315-A, 80.402-A or 80.402-B), then the space must be rated as a chemical storage or use room and classified as either H-1, H-2, H-3, H-6 or H-7 occupancy. Tables 9-A and 9-B of the UBC combine these various tables to form a more comprehensive overall "exempt amounts" listing, but 9-A and 9-B are somewhat harder to read due to the potential combinations of footnotes (the increases allowed by each footnote may be compounded). Either the UFC tables or the UBC tables may be used, as the values are the same.

3.6   HAZARDOUS MATERIAL STORAGE ROOMS

3.6.1   When the quantities of hazardous materials to be stored or used exceed the amounts allowed in a control area (i.e., exceed the exempt amounts given in Tables 9-A or 9-B), a hazardous occupancy must be created.

UBC Chapter 9 identifies four distinct divisions within Group H which cover hazardous material storage rooms.

UBC 901(a) Requirements for Group H Occupancies - Group H Occupancies Defined

"Division 1. Occupancies with a quantity of material in the building in excess of those listed in Table 9-A which present a <u>high explosion hazard</u>, including, but not limited to:

1.   Explosives, blasting agents, fireworks and black powder . . .
2.   Unclassified detonatable organic peroxides.
3.   Class IV oxidizers.
4.   Class IV or Class III detonatable unstable (reactive) materials."

This type of storage <u>must</u> be in a dedicated building, with no other occupancy.

"Division 2. Occupancies with a quantity of material in the building in excess of those listed in Table No. 9-A which present a moderate explosion hazard or a hazard from accelerated burning, including, but not limited to:

1.   Class I organic peroxides.

2.   Class III nondetonatable unstable (reactive) materials.

3.   Pyrophoric gases.

4.   Flammable or oxidizing gases.

5.   Class I, II or III-A flammable or combustible liquids which are used in normally open containers or systems or in closed containers pressurized at more than 15-pounds-per-square-inch gauge. . .

6.   Combustible ducts in suspension or capable of being put into suspension in the atmosphere of the room or area . . .

7.   Class III oxidizers."

"Division 3. Occupancies with a quantity of material in the building in excess of those listed in Table No. 9-A which present a high fire or physical hazard, including, but not limited to:

1.   Class II, III or IV organic peroxides.

2.   Class I or II oxidizers.

3.   Class I, II or III-A flammable liquids or combustible liquids which are utilized or stored in normally closed containers or systems and containers pressurized at 15-pounds-per-square-inch gauge or less and aerosols.

4.   Class III-B combustible liquids.

5.   Pyrophoric liquids or solids.

6.   Water reactives.

7.   Flammable solids, including combustible fibers or dusts, except for dusts included in Division 2.

8.   Flammable or oxidizing cryogenic fluids (other than inert).

9.   Class I unstable (reactive) gas or Class II unstable (reactive) materials."

"Division 7. Occupancies having quantities of materials in excess of those listed in Table No. 9-B that are <u>health hazards</u>, including:

1.   Corrosives.
2.   Toxic and highly toxic materials.
3.   Irritants.
4.   Sensitizers.
5.   Other health hazards."

(Note: It should be pointed out that while a significant amount of corrosives, sensitizers and irritants are permitted before an H-7 occupancy is mandated, only very small amounts of toxic solids, liquids or gases are tolerated in a non-H-7 occupancy.)

Each of the individual storage occupancies has distinct requirements for construction and operation established by the codes. Article 80 of the UFC provides specific information regarding H-7 occupancies; Article 79 provides specific information regarding H-3 occupancies.

The 1993 edition of the BOCA National Code, which may not yet be adopted by many jurisdictions, incorporates provisions for "exempt" amounts of various hazardous materials much like the UBC Tables 9-A and 9-B. BOCA provides four "hazardous use groups" -- H-1, H-2, H-3 and H-4. The H-1, H-2 and H-3 groups are similar to the same UBC divisions, while H-4 is very similar to UBC H-7.

In the special case of an occupancy involved in the manufacturing of semiconductor components (or other similar research and development labs) with more than exempt amounts of hazardous materials, a Group H, Division 6 occupancy is mandated. In this case, UBC 911(e) requires the hazardous materials be placed in an inside liquid storage room (in compliance with 901(d) for Class I, II or III-A, flammable or combustible liquids) or an HPM storage room not to exceed 6,000 square feet in area.

3.6.2   Flammable Liquid Storage:

UBC Chapter 9 addresses specific requirements for flammable liquid use and storage rooms in Sections 901(c) and 901(d) as follows:

UBC 901(c)1-5   Group H Occupancies Defined - Liquid Use, Dispensing and Mixing Rooms

"Rooms in which Class I, Class II and Class III-A flammable or combustible liquids <u>are used, dispensed or mixed in open containers</u> shall be constructed in accordance with the requirements for a <u>Group H, Division 2 Occupancy</u> and the following:

1   Rooms in excess of 500 square feet shall have at least one exterior door approved for fire department access.

2   Rooms shall not exceed 1000 square feet in area.

3   Rooms shall be separated from other areas by an occupancy separation having a fire-resistive rating of not less than one hour for rooms up to 150 square feet in area and not less than two hours where the room is more than 150 square feet in area. Separations from other occupancies shall not be less than required by Chapter 5, Table No. 5-B.

4   Shelving, racks and wainscoting in such areas shall be of noncombustible construction or wood not less than one inch in nominal thickness.

5   Liquid use, dispensing and mixing rooms shall not be located in basements."

UBC 901(d)1-4   Group H Occupancies Defined - Liquid Storage Rooms

"Rooms in which Class I, Class II and Class III-A flammable or combustible liquids are <u>stored in closed containers</u> shall be constructed in accordance with the requirements for a <u>Group H, Division 3 Occupancy</u> and adhere to the following:

1   Rooms in excess of 500 square feet shall have at least one exterior door approved for fire department access.

2   Rooms shall be separated from other areas by an occupancy separation having a fire-resistive rating of not less than one hour for rooms up to 150 square feet in area and not less than two hours where the room is more than 150 square feet in area. Separations from other occupancies shall not be less than required by Chapter 5, Table 5-B.

3   Shelving, racks and wainscoting in such areas shall be of noncombustible construction or wood of not less than one inch nominal thickness.

4   Rooms used for the storage of Class I flammable liquids shall not be located in a basement."

(The previous H-3 code requirement of a maximum 1000-square-foot room size now applies only to H-2 dispensing rooms.)

3.6.3   Flammable/Combustible Storage and Dispense Rooms:

When the quantities of Class I, II and III-A flammable or combustible liquids exceed those stipulated in UBC Table 9-A, special storage or dispense/mixing rooms shall be constructed to comply with the requirements of UBC Section 901(c), (d) or (e) and 902. Where the size of an HPM storage room is such that it must have at least one outside wall (more than 500 square feet), the outside wall shall not be less than 30 feet from the property line (or other buildings).

UBC 901 and UFC Article 80 stipulates the type of storage facility -- H-2 or H-3 -- required, depending on whether the material is merely stored (H-3) or it is used, dispensed or mixed (H-2).

3.6.4   H-6 Occupancy Storage of HPMs:

UBC 911(e) requires that storage of HPMs, in excess of that listed in Tables 9-A or 9-B, be in inside rooms as outlined in 901(d) or in HPM storage rooms, the area of which shall not exceed the following:

- <u>For Storage</u>:
  not in excess of 6,000 square feet
  - if greater than 300 square feet must be two-hour or greater fire-resistive.
  - if less than ?00 square feet must be one-hour or greater fire-resistive.
  - located on outside wall if greater than 500 SF.

- <u>For Dispensing</u> of Class I or II flammable liquids or flammable gases:
  not in excess of 1,000 square feet
  - located on outside wall if greater than 500 SF.

3.6.5   UFC Article 51 addresses the requirements for outside storage of HPMs (Section 51.110(a); see also UFC 80.301(aa)) and storage of HPMs within buildings, Section 51.110(b). For the purpose of this document, we are primarily concerned with inside storage rooms. In general, 51.110(b) addresses the following issues relative to <u>H-6</u> storage facilities:

- Allowable area.
- Fire rating of construction.
- Location within the building, i.e., on outside wall or not.
- Separation of different chemical types.
- Maximum quantity of material stored.
- Ventilation requirements.
- Emergency alarms and annunciation.
- Drainage and containment.
- Gas detection, annunciation and mitigation.
- Electrical hazard rating.

3.6.6   Separation of HPMs:

Not only does an HPM storage room need to be separated with fire-resistive construction from the rest of the building and other occupancies, but <u>areas within the given storage room</u> used for differing classifications of HPMs -- toxics, acids, flammables, oxidizers, etc. -- <u>must also be separated</u>. Table 51.110-A of the UFC defines the material separation requirements. Additional requirements are stipulated in Article 80 for specific HPMs.

UFC Table 51.110-A requires separation of the following HPMs with either one-hour or noncombustible partitions, except that water reactives and flammables may not be in the same storage room. Where toxic materials are stored, they must be separated from <u>all</u> other categories by one-hour-rated construction.   Generally the requirement for separation of the various materials (excluding toxics) is a non-combustible partition which extends at least 18" above and to the front and rear of the stored material.

- Toxics.
- Acids.
- Bases.
- Flammables.
- Oxidizers.
- Water Reactives.
- Pyrophorics.

3.6.7   Article 79 of the UFC addresses requirements for <u>flammable and combustible liquids</u> in general.   Liquid storage rooms for these materials are specifically addressed in UFC 79.203.   <u>Inside</u> liquid storage rooms have been deleted from the 1991 edition of the UFC.

Liquid storage rooms are classified as H-3 and (per the UBC) must have at least one outside wall.   The outside wall must meet the requirements for distance from a property line or an adjacent building.

UFC Article 79 covers storage quantities and arrangements, sprinkler requirements, spill and drainage control, and ventilation requirements.

3.6.8   **Division III of Article 80** prescribes very specific requirements for the <u>storage</u> of various categories of hazardous materials. Section 80.301 outlines the general requirements for <u>storage</u> of materials in containers, cylinders and tanks in excess of the exempt amounts given throughout the material specific sections of Division III. *(Further discussion of the technical issues is provided in Chapters 2, 4, 5, 6 and 7 of this book.)*

3.6.9   **Ventilation Systems:**

Mechanical exhaust ventilation for HPM storage/use areas should be independent from other areas. *(Design requirements are outlined in Chapter 4.)* UFC Article 80 requires that emergency power, in lieu of standby power, be used for the exhaust ventilation (including the power supply for treatment systems) when toxic or highly toxic compressed gases are stored or dispensed. *(The distinction between emergency power and standby power for ventilation systems is a subtle one and the practicalities, or lack thereof, of utilizing standby power are discussed in Chapter 6.)*

3.6.10   **Temperature Control:**

Temperature-control systems are required when outside ambient conditions are not suitable for the material being stored (UFC 80.301(t)3. A redundant temperature control system must be provided to prevent a hazardous reaction upon failure of the primary control system.

3.6.11 Piping System:

Piping for HPM must be appropriate to the material being handled at the conditions of pressure, temperature, etc. relevant to use (see UFC 80.301(c)).   Additional requirements are stipulated for toxic and highly toxic compressed gases.   (*Refer to Section 3.8 for a more detailed discussion of piping requirements.*)

3.6.12 Fire Suppression:

(*The requirements for fire suppression systems within chemical storage and/or dispensing/use facilities are outlined in Chapter 5.*) An important design requirement related to the sprinkler system is the need for drainage and containment of the water discharged by the sprinkler system for certain occupancies, as defined in Articles 51, 79, and 80.

UFC 79.115(c)1 Spill Control, Drainage Control and Secondary Containment
"When drainage control is required, rooms, buildings or areas shall be provided with a drainage system to direct the flow of liquids to an approved location or treatment system, or shall be provided with secondary containment for the flammable or combustible liquids and fire-protection water."

UFC 79.115(c)4(ii) Spill Control, Drainage Control and Secondary Containment - Neutralizers and Treatment Systems
"Drainage systems... shall comply with the following...
(ii)   Overflow control from the neutralizer or treatment system shall be provided to direct liquid leakage and fire-protection water to a safe location away from buildings, material or fire-protection control valves, means of egress, adjoining properties, or fire department access roadways."

Facilities needed for such spill and drainage control may include underground or above ground tanks, pits, vaults or sumps.

3.6.13   Toxic Gas Monitoring:

A continuous gas detection system must be provided to detect the presence of hazardous gases at or below the permissible exposure limit (PEL) or the ceiling limit (UFC 80.303(a)9).   The detection system must initiate a local audible and visual alarm both inside and outside the storage area with the audible alarm being distinct from all other alarms as well as transmitting a signal to a constantly attended plant Emergency Control Station (ECS).

Gas detectors must be placed within the room or area and also at the discharge from the ventilation treatment system.   Detection of hazardous gases at 1/2 the IDLH limit or more at the ventilation treatment discharge shall initiate an alarm.   Gas cabinets dispensing compressed gases must also contain gas detectors and must be internally sprinklered.   Activation of the alarm/monitoring system must automatically close the shut-off valve on highly toxic or toxic gas supply lines related to the system being monitored.

(Note: Some recent interpretations of the Americans with Disabilities Act (ADA) may require use of new types of strobes, as previous visual alarms may not be acceptable.)

(More on this subject is provided in Chapter 7.)

3.6.14   Smoke Detection:

Supervised smoke detection must be provided in rooms or areas where highly toxic compressed gases are stored indoors.   Activation of the detection systems shall sound a local alarm and notify the ECS (UFC 80.303(a)10).   An approved emergency signal device is also required to be installed outside of each interior exit door of storage buildings, rooms or areas.   A dedicated telephone system which allows personnel to inform the ECS of the exact nature and severity of the condition is recommended. Activation of the manual alarm must sound a local alarm as well as send a signal to the ECS (UFC 80.301(u) and (v)).

(More on this subject is provided in Chapter 7.)

3.6.15   Pyrophoric Materials:

Requirements for pyrophoric materials storage are covered in UFC Section 80.308. Storage of pyrophoric gases (such as silane and certain mixtures of phosphine) require that electrical wiring and equipment in storage areas comply with NEC Class 1, Division 2 installations (UFC 80.308(a)6). This, however, conflicts with the provisions of NEC 500-2, which states that areas used exclusively for pyrophoric materials need not be classified. The rationale behind this NEC statement is that, since pyrophoric materials need no ignition source to burn (i.e., they are self-igniting), the absence of ignition sources provided by Class I, Division 2 wiring and equipment does not provide any additional safeguard. This rationale has been somewhat accepted, in that Class 1, Division 2 wiring and equipment are no longer required for storage areas for pyrophoric liquids. The local Authorities Having Jurisdiction over the project should provide written clarification on this issue. *(Explosion-proof installations are expensive to install and maintain and are discussed further in Chapter 6.)*

As mentioned in Chapter 1, there is a proposed ordinance pertaining specifically to silane tentatively named UFC Article 80-1. Wafer manufacturing facilities commonly utilize this pyrophoric gas (SiH_4) as a source of pure silicon. Article 80-1 would apply to installations using silane or gas mixtures with more than 2 percent silane, where the quantity of such gas is more than 100 cubic feet (at STP) and the gas is in sprinklered facilities, in gas cabinets.

Some recommended practices with respect to the storage and dispensing of pyrophoric materials include:

- Restricted flow orifice (0.010") in the gas cylinder.
- Automatic sprinkler system.
- Ventilation at dispensing fittings subject to leaks, such that velocity is not less than 200 feet per minute.
- Separation of storage.
- Standby electrical power.

The objectives of these recommendations are:

1. Reduce the potential for detonation.
2. Reduce the potential for fast burning (deflagration).
3. Reduce the impact on the area in which silane is stored from the pressures created by silane "reaction."

3.7   HAZARDOUS MATERIAL DISPENSING, USE AND HANDLING

3.7.1   **Article 80 Division IV** addresses the dispensing, use and handling of hazardous materials.   In addition, UFC Article 79 Division VIII covers the use, dispensing and mixing of Class I, II or II-A liquids both inside and outside of buildings.   In general, UFC 79.805 requires such use of flammables or combustibles in an H-2 or H-3 Occupancy.

The use of bulk chemical dispensing and distribution systems is becoming more prevalent in facilities with high consumption rates of hazardous materials.   There are many good technical reasons for the use of bulk dispense and distribution, such as:

- Assured High Purity at Point-of-Use
- Continuous supply not reliant on frequent deliveries
- Reduced hazard due to handling outside the occupied facility

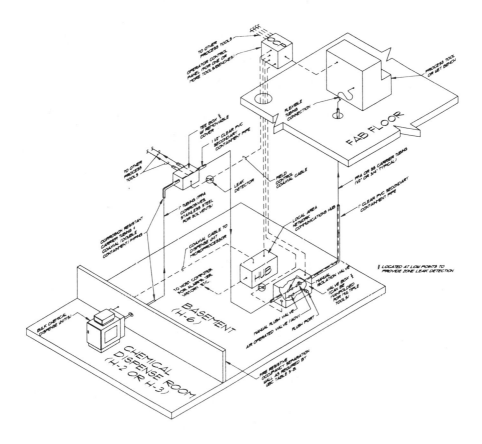

Figure 3.3 BULK CHEMICAL DISPENSE SCHEMATIC Illustrating Relationship of Various Components to the Process Tool. (Illustration courtesy of Systems Chemistry, Inc.)

- Reduced hazard due to less frequent handling of larger containers

In addition, the storage of chemicals outside the occupied area reduces the inventory of HPMs and may help to bring a facility with excessive inventories into code compliance (remember, the quantity of HPM in the piping is not counted in the limitations imposed by UFC Article 51). (*Refer to* **Figure 3.3** *for a schematic illustration of how such a bulk dispense and distribution system functions.*)

The manner in which an HPM is dispensed, handled or used shall not increase its overall hazard to life or property unless such increased hazards are mitigated in an approved manner. (*Refer to* **Figures 3.4 and 3.5** *for illustrations of HPM dispense facilities and components.*)

3.7.2   Hazardous materials dispensing, use and handling requirements are more specifically addressed in the following code excerpts:

UFC 80.402(b)1   Hazardous Materials - Indoor Dispensing and Use
"Indoor dispensing and use of hazardous materials shall be in accordance with the provisions of this subsection and Section 80.401."

UFC 80.401(c)1 Dispensing, Use and Handling - Piping, Tubing, Valves and Fittings
"Piping, Tubing, Valves and Fittings . . . "

UFC 80.401(c)2C (Dispensing, Use and Handling) Piping, Tubing, Valves and Fittings - Design and Construction
"Emergency shutoff valves shall be identified, and the location shall be clearly visible . . . "

Figure 3.4 AUTOMATIC CHEMICAL DELIVERY UNITS Installed in HPM Dispense Room. (Photo courtesy of Systems Chemistry, Inc.)

Figure 3.5 CHEMICAL DISPENSE UNIT For Multiple User Distribution System. (Photo courtesy of Systems Chemistry, Inc.)

UFC 80.401(c)3C (Dispensing, Use and Handling) Piping, Tubing, Valves and Fittings - Supply Piping
"Where gases of liquids are carried in pressurized piping above 15 psig, excess flow control shall be provided . . . "

UFC 80.401(c)3D(i-ii) (Dispensing, Use and Handling) Piping, Tubing, Valves and Fittings - Supply Piping
"Readily accessible manual or automatic remotely-activated fail-safe emergency shut-off valves shall be installed on supply piping . . .
(i)   The point of use, and
(ii)   the tank, cylinder or bulk source"

3.7.3   *Gas and smoke detection systems as described in Section 3.6 and Chapter 7 are also required for indoor dispensing and use of HPMs.*

3.7.4   **Special requirements for toxic and highly toxic gases:**

UFC 80.401(c)3A (Dispensing, Use and Handling) Piping, Tubing, Valves and Fittings - Supply Piping
"Piping and tubing utilized for... highly toxic or toxic material shall have welded or brazed connections throughout unless an exhausted enclosure is provided if the material is a gas...
            Exception: Nonmetallic piping with approved connections."

UFC 80.303(a)6D(i,iii)   (Toxic and Highly Toxic Compressed Gases) Ventilation and Storage Arrangement - Treatment Systems
(i)   "Treatment systems shall be utilized to handle the accidental release of gas. Treatment systems shall be utilized to process all exhaust ventilation to be discharged from gas cabinets, exhausted enclosures or separate storage rooms...

(iii)   **Performance.** Treatment systems shall be designed to reduce the maximum allowable discharge concentration of the gas to one-half IDLH at the point of discharge to the atmosphere...."

UFC 80.402(b)3G(i,iv) (Indoor Dispensing and Use) Closed Systems - Special Requirements

(i)   "**Ventilation and Storage Arrangement.** Compressed gas cylinders in use shall be within ventilated gas cabinets, laboratory fume hoods, exhausted enclosures or separate gas storage rooms...

(iv)   **Treatment Systems.** Treatment systems shall be provided in accordance with Section 80.303(a)6D."

Ventilation, treatment and gas detection requirements apply for both indoor and outdoor dispensing and/or use sites.   See UFC 80.402(c)B, C and D.

UFC 80.402(b)3G (Indoor Dispensing and Use) Closed Systems - Special Requirements

(viii)   "**Process equipment.** Effluent from process equipment containing highly toxic or toxic gases which could be discharged to the atmosphere shall be processed through an exhaust scrubber or other processing system.   Such systems shall be in accordance with the Mechanical Code as required for product conveying ventilation systems."

(Refer to Chapter 4 for a detailed discussion of treatment system requirements.)

The storage and dispensing of hazardous gases (those with flammability, health hazard or reactivity rating of 3 or 4) in gas cabinets, exhausted enclosures or within a "work station" is mandated by UFC 51.107(a) for an H-6 occupancy. In other "H" occupancies, the indoor storage of toxic or highly toxic gases shall be within ventilated gas cabinets, exhausted enclosures or within a ventilated separate gas storage room (UFC 80.303(a)6). Pyrophoric gases in excess of 50 CF must be in gas cabinets (UFC 80.308(a)2). The use or dispensing of flammable, toxic or highly toxic gases shall be within ventilated gas cabinets, laboratory fume hoods or exhausted enclosures (UFC 80.402(b)2C and UFC 80.402(b)3G).

It is recommended that any gases with health hazard, flammability or reactivity ranking of 3 or 4 be used or dispensed in a facility with the cylinder and any fittings subject to leakage in a gas cabinet with suitable ventilation and controls. *The requirements for gas cabinet ventilation and life safety systems are outlined in Chapters 4 and 7.* Modern gas cabinets come with a variety of control and safety features designed to make the operation and changing of cylinders as safe as possible. *(Refer to* **Figures 3.6, 3.7, 3.8 and 3.9** *for illustrations of HPM gas cabinets and their features.)*

As discussed earlier, several jurisdictions have enacted Toxic Gas Ordinances (TGO) or Toxic Gas Model Ordinances (TGMO). In the event your jurisdiction has such an ordinance, it may require that you comply with current UBC and UFC requirements for the storage, handling and treatment of gas releases, whether or not you modify or alter your facility.

3.8   HPM PIPING SYSTEMS

3.8.1   General:

A complete discussion of the requirements for HPM piping systems is beyond the scope of this book. However, in the interest of providing some guidelines and direction, the following is offered. For brevity, code excerpts are often paraphrased; the reader should consult the actual code text for exact wording. The intent, however, has not been changed. Commentary remarks are found in parentheses. Where requirements differ, resolution of discrepancies can be met by meeting the most stringent requirements.

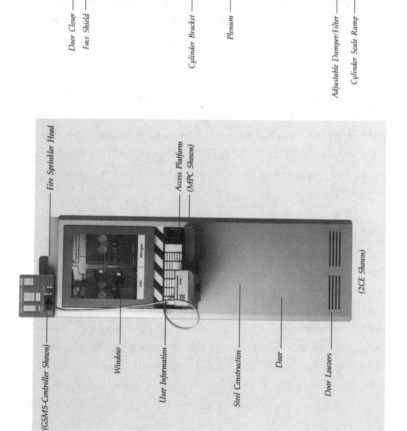

Figure 3.7 INTERNAL VIEW OF GAS CABINET Illustrating Piping, Ventilation and Safety Components. (Illustration courtesy of Semi Gas Systems)

Figure 3.6 EXTERNAL VIEW OF GAS CABINET. (Illustration courtesy of Semi Gas Systems)

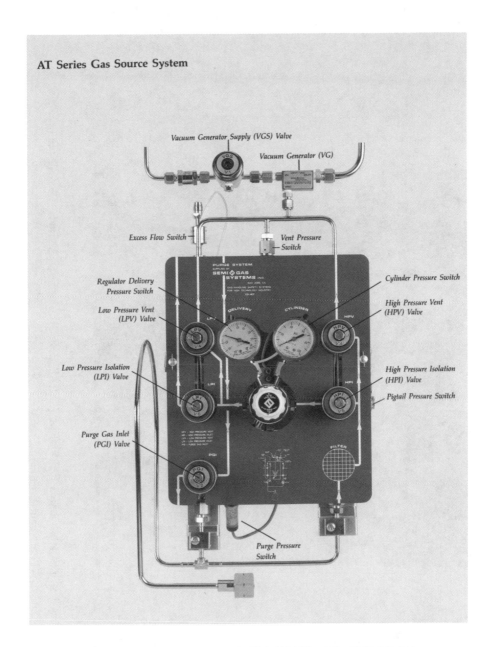

Figure 3.8 DETAILS OF AUTOMATIC VENT AND PURGE SYSTEM For Hazardous Gas Cylinder Enclosure. (Photo courtesy of Semi Gas Systems)

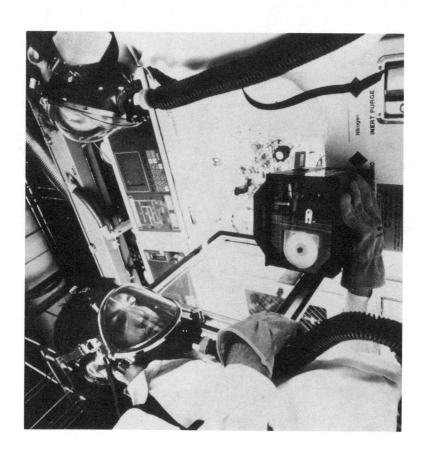

Figure 3.9 TECHNICIANS PERFORM A LEAK CHECK of a Toxic Cylinder Gas. (Note Breathing Apparatus.) (Photo courtesy of Air Products)

3.8.2   H-6 Occupancy General Requirements (UBC 911(f)) Piping:

Piping shall be installed in accordance with nationally recognized standards. (An example of this would be the ANSI B31.3 Chemical Plant and Petroleum Refinery Piping.)

Piping and tubing shall be metallic unless the material being conveyed is incompatible. (Most corrosive liquids which are bulk dispensed are piped in teflon tubing or other chemically inert plastic and are double contained.)

Systems supplying gaseous HPMs which have a health hazard ranking of 3 or 4 shall be <u>welded throughout</u>, except for connections, valves, fittings, etc. which are within ventilated enclosures. (*Refer to* **Figure 3.10** *for illustration.*) Where highly toxic and pyrophoric gases are installed inside the building, coaxial tubing is recommended. (*Refer to* **Figure 3.11**). In general, Type 316 electropolished stainless steel tubing is utilized for HPM gases; however, some of the chlorine and bromine containing corrosive gases are better handled in Hastelloy C-22 tubing, due to its greater resistance to corrosion. Because Hastelloy C-22 is relatively high in cost, a popular alternative to C-22 is vacuum arc remelt (or VAR) 316 stainless steel. Under this process, the tubing stock is re-melted under a vacuum to reduce contaminant levels. With a substantially lower sulfur level in the tubing, sulfur "stringers" (and the possibility for corrosion along them) are much reduced, resulting in improved corrosion-resistance. *As an example of material selections for typical high-purity semiconductor gases, we have included a* **Schedule of Materials as Table 3.3.** *(NOTE: This schedule is provided for information only; each user must decide on the optimum materials.)*

HPM supply piping in service corridors shall be exposed to view (to facilitate servicing and visual observation of condition at all times).

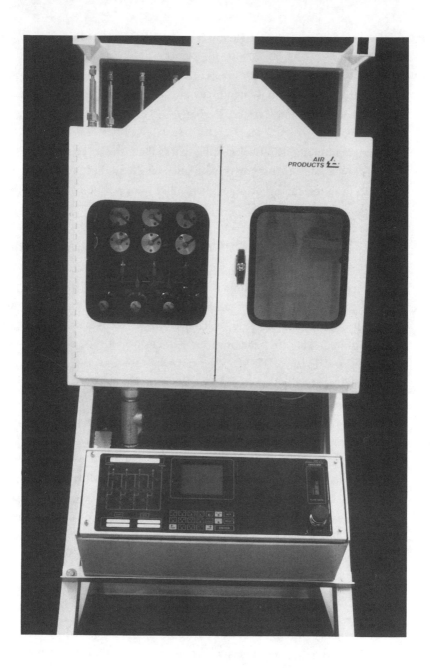

Figure 3.10 VALVE MANIFOLD BOX For Hazardous Gas Distribution to Multiple Points-of-Use. (Photo courtesy of Air Products)

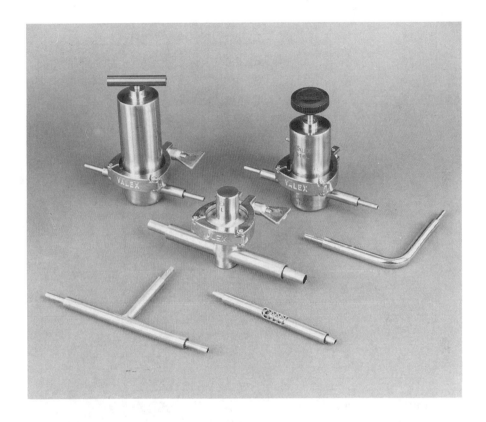

Figure 3.11   COAXIAL TUBING,   VALVES AND FITTINGS for Conveyance of Toxic and Hazardous Process Gases. (Photo courtesy of Valex Corp.)

ULTRAPURE PIPING SCHEDULE / Specialty Gas	GAS CLASSIFICATION					PIPE MATERIAL		
	Oxidizer	Flammable	Toxic	Corrosive	Pyrophoric	SS Coax Tube	SS Single Wall	Hastelloy C-22
Ammonia		▓	▓				▓	
Argon							▓	
Arsine		▓	▓			▓		
Boron Trichloride			▓	▓				▓
Boron Trifluoride			▓	▓		▓		
Chlorine			▓	▓				▓
Diborane & Diborane Mixtures		▓	▓		▓	▓		
Dichlorosilane		▓	▓	▓	▓	▓		
Halocarbon 14							▓	
Halocarbon 23							▓	
Halocarbon 116							▓	
Helium							▓	
Hydrogen		▓					▓	
Hydrogen Bromide			▓	▓				▓
Hydrogen Chloride			▓	▓				▓
Hydrogen Fluoride			▓	▓		▓		
Hydrogen/Nitrogen Mixtures		▓					▓	
Nitrogen							▓	
Nitrogen Trifluoride	▓		▓				▓	
Nitrous Oxide	▓						▓	
Oxygen	▓						▓	
Phosphine & Phosphine Mixtures		▓	▓		▓	▓		
Silane		▓			▓	▓		
Sulfur Hexafluoride							▓	

Note: For reference and information only; each facility and installation must select the proper materials.

Table 3.3 HPM GAS PIPING CONSIDERATIONS.

HPM piping installed <u>above</u> exit corridors or other occupancies (not H-6) must meet the following requirements:

- Automatic sprinklers.
- Ventilate at 6 air changes per hour (minimum).
- Receptor to collect liquids, i.e., double containment.
- One-hour fire-resistive separation.
- Accessible manual or automatic shut-off valves, except on waste lines.
- Excess flow control valves. See UFC 51 below.
- Class I, Division 2 Electrical.

All piping shall be identified in accordance with recognized standards.

3.8.3   **UFC Article 51 Semiconductor Fabrication Facilities Using HPMs - Provisions (for H-6 Occupancies):**

UFC 51.105(e)1   (Fabrication Areas) Special Provisions - Excess Flow Control
"Where HPM supply gas is carried in pressurized piping, a fail-safe system shall shut off flow due to a rupture in the piping. . ."

UFC 51.105(e)2   (Fabrication Areas) Special Provisions - Piping and Tubing Installation
"Piping and tubing shall be installed in accordance with approved standards. Supply piping for HPM having a health hazard ranking of 3 or 4 shall have welded connections throughout unless an exhausted enclosure is provided. . ."

UFC 51.106(f)1A-C   (Workstations within Fabrication Areas) Special Provisions - Chemical Drainage and Containment
Each workstation utilizing HPM liquids shall have:
A   Drainage piping to compatible system.

B   Work surface sloped to direct spills to containment system.

C   Means to contain or direct spills to the drainage system.

UFC 51.106(f)2 (Workstations within Fabrication Areas) Special Provisions - Identification

"...Labels shall be affixed so as to be conspicuously visible at all times."

UFC 51.106(f)3 (Workstations within Fabrication Areas) Special Provisions - Shut off Valves

Readily accessible, located at work station and per UFC 51.108(d)2.

UFC 51.108(d)2-5   Handling of HPM within Exit Corridors - Piping

2   Shut off valves.

3   Excess flow control (per 51.105(e)1).

4   Gas detection (per 51.105(e)3).

5   Electrical in piping space Class I, Division 2.

UFC 51.109(e) Handling of HPM within Service Corridors - Piping

Same as fab; see 51.105(e).

UFC 51.110(a) Storage of HPM - Outside Storage

Emergency shutoffs in HPM gas piping per 51.105(e)1. Manual emergency shut off valve located outside building for each HPM pipe from outside storage.

3.8.4   UFC Article 79 Division VII Piping, Valves and Fittings:

79.701(a) Materials and Designs.
Piping materials and components shall be designed and fabricated from suitable material of adequate strength and durability, in accordance with engineering standards, listed for the application, or approved.

79.701(b) Low Melting Point Materials.
Low melting point materials -- aluminum, copper and brass -- shall be protected against fire exposure.

79.702 Protection from Corrosion and Galvanic Action.
Piping shall be non-corroding or protected from corrosion. Dissimilar metal parts may not be connected.

79.703 Valves.
Valves shall be provided to permit proper operation and isolation and to protect the plant.

79.704 Supports.
Piping systems shall be substantially supported.

Discussion:

Allowance should be made for thermal expansion/contraction, earthquake, etc.

79.706 Pipe Joints.
Joints in piping shall be liquid-tight, welded, flanged or threaded.

Discussion:

Compression or friction fittings use is allowed only in certain locations; see code text.

79.707 Bends.

Bends in pipe and tubing shall be not in excess of 90 degrees or at a radius of less than five diameters to the inside edge.

Discussion:

Although this code requirement applies only to flammable/combustible liquid piping, recommended practice is that this minimum apply to all HPMs. This concept is particularly important for ultrapure stainless steel tubing used to convey corrosive gases, where this tubing is frequently field bent to save welding.   Bending to less than ten diameters may result in accelerated corrosion at stress riser cracks, leading to eventual failure of the tubing.

79.708 Testing.

Test all piping to 150 percent of the maximum pressure subjected to, but not less than, 5 psig at the highest point, for at least 10 minutes.

3.8.5   UFC Article 80 Division III Storage:

80.301(c) Piping, Valves and Fittings.

Adequate strength and compatibility.

80.303(a)8 (Toxic and Highly Toxic Compressed Gases) Indoor Storage - Limit Controls

Excess flow control for stationary tanks.

80.303(b)4A-C   (Toxic and Highly Toxic Compressed Gases) Exterior Storage - Piping and Controls

A   Pressure relief valves vented to treatment system.

B   Local exhaust at fill/dispense connections, directed to treatment system.

C   Excess flow control for inlet and outlet connections of stationary tanks.

3.8.6   **UFC Article 80 Division IV Dispensing, Use and Handling:**

80.401(c)2A-D   Piping, Tubing, Valves and Fittings - Design and Construction
A   Compatible with the material.
B   Identified per recognized standards.
C   Emergency shutoff valves identified and located.
D   Backflow prevention or check valves provided to prevent a hazardous condition or discharge.

80.401(c)3A-D Piping, Tubing, Valves and Fittings - Supply Piping
A   Welded or brazed connections, unless an exhausted enclosure (for gases) or receptor (for liquids) is provided.
B   Not in exit corridors or other than H occupancies.
C   Excess flow control above 15 psig.
D   Accessible manual or automatic fail-safe emergency shutoff valves at point of use and source.

3.9   CODE CITATIONS

In the interest of providing a quick-reference guideline for the reader, we offer a compilation of the applicable UBC and UFC Sections concerning HPM storage occupancies. The requirements include, <u>but may not be limited to</u> the following:

Construction:
-      UBC 503
-      UBC 505
-      UBC 506
-      UBC 507
-      UBC 901
-      UBC 902
-      UBC 911(e)
-      UFC 51.110(b)
-      UFC 79.202

Construction: (Cont'd.)
- UFC 79.203
- UFC 79.804
- UFC 80.105

Spill Containment:
- UBC 902(b), (c), (d), (e)
- UFC 51.110(b)7
- UFC 79.115
- UFC 79.805(k)
- UFC 80.301(l)
- UFC 80.401(g)

Location:
- UBC 903 (on property)
- UBC 911(e)2 (within building)
- UFC 51.110(b)2 (within building)

Exiting:
- UBC 911(e)3
- UBC 3319

Explosion Venting:
- UBC 910
- UFC 79.804.5 (dispensing and mixing)
- UFC 51.110
- UFC 80.301(q)
- UFC 80.303(a)4
- UFC 80.305(a)4
- UFC 80.402(b)2E
- NFPA Article 68 (Guide for Venting of Deflagrations)

Separation of Chemical Types:
- UFC 51.110(b)3
- UFC Table 51.110-A
- UFC 80.301(n)
- UFC 80.303(a)6C
- UFC 80.401(e)

Piping Systems:
-       UBC 911(f)
-       UFC 51.105(e)
-       UFC 51.106(f)
-       UFC 51.108(d)
-       UFC 51.110(a)2
-       UFC 79.701 through 79.708
-       UFC 80.301(c)
-       UFC 80.303(a)8 (indoor gas storage)
-       UFC 80.303(b)4 (exterior gas storage)
-       UFC 80.401(c) (handling, dispensing)

Ventilation Requirements:
-       (*See Chapter 4 of this book.*)

Fire Suppression:
-       (*See Chapter 5 of this book.*)

Electrical Requirements:
-       (*See Chapters 6 and 7 of this book.*)

SUMMARY

HAZARDOUS MATERIALS STORAGE AND HANDLING (3.0)

◆   As each facility and project is unique, the methods of providing code compliance will necessarily be unique to each project.

◆   Design issues related to Heating, Ventilating and Air Conditioning, Fire Suppression, Power and Alarm/Monitoring are discussed in Chapters 4, 5, 6 and 7, respectively.

HAZARDOUS MATERIALS (3.2)

◆   HAZARDOUS PRODUCTION MATERIAL (HPM) is a solid, liquid or gas that has a degree-of-hazard rating in health, flammability or reactivity of Class 3 or 4 as ranked by UFC standard No. 79-3 and which is used directly in research, laboratory or production processes which have as their end-product materials which are not hazardous. (UFC 9.110)

◆   An HPM STORAGE ROOM is a room used for the storage or dispensing of HPMs and which is classified as a Group H, Division 2, 3 or 7 Occupancy.

◆   Classification of HPMs: For guidance on the classification of chemicals by hazard category, refer to Appendix VI-A of the UFC. For additional assistance, the supplier of each material should be consulted.   Manufacturers are required to publish "Material Safety Data Sheets" (MSDS).

◆   Article 80 of the UFC provides general requirements for all types of hazardous materials regardless of the occupancy or use of a facility.

◆   Article 80 subdivides hazardous materials classifications into two broad categories:

   ▪   <u>Physical Hazards</u>
   ▪   <u>Health Hazards</u>

◆   Materials with a primary classification as a physical hazard may also present a health hazard, and materials with a primary classification as a health hazard may also present a physical hazard. <u>All of the hazards must be addressed</u> by the facility design (see UFC Section 80.101(c)).

MAXIMUM ALLOWABLE QUANTITIES OF HPMS (3.3)

◆   <u>Absolute Quantity</u>:   UFC Table 51.105B stipulates the maximum permissible quantities (in gallons, pounds, or cubic feet) of various categories of HPMs which may be contained within a given H-6 occupancy, regardless of the size of the facility.

◆   <u>Density Limit</u>:   UFC Table 51.105A limits the density in lb./SF, gal./SF or CF/SF of HPM solids, liquids and gases which may be stored, used or dispensed in a single H-6 fabrication area.

◆   In-Use within Workstations: The quantities of chemicals in-use within a workstation are governed by Table 51.106-A in UFC 51.106(b).

◆   Piped Systems: Quantities of HPMs in utility, process or waste pipes are not included in quantity limitations.

◆   Unsegregated Storage within Fab: UFC Section 51.107 regulates HPMs within the fab stored outside of workstations.

◆   Segregated Storage (HPM Storage Rooms) - within the H-6:   UBC 911(e)1 states that stored quantities of HPMs exceeding the maximum amounts listed in UBC Tables 9-A or 9-B must be in a room complying with UBC requirements for an inside liquid storage room or an HPM storage room.

◆   Storage rooms provide separation, protection, size limitation and location such that the stored HPM is not an additional hazard to the fab.

TRANSPORTATION OF HPMs (3.4)

◆   Manual Transportation: Manual transportation of HPMs within a single fab area is not specifically limited other than requiring the use of approved chemical containers.

◆   Manual transportation of HPMs between fabs, or between segregated storage rooms, however, requires the creation of a service corridor.

◆   UFC 51.108(a) stipulates that HPMs are not permitted in exit corridors of new buildings.

◆   In existing buildings where the exit corridors must be used for transportation of HPMs, this practice may be continued, provided the owner complies with the requirements of UFC 51.108(b) and UBC 911(c).

CONTROL AREAS (3.5)

◆   Areas within a facility which use or store less than the "exempt" amounts of HPMs may be considered "control areas" and must be constructed with not less than one-hour partitions.

◆   The "control area" is a space within an occupancy which does not have to be considered a specific hazardous storage occupancy such as H-1, H-2, H-3 or H-7.

◆   There may be up to four individual control areas within a building.

HAZARDOUS MATERIAL STORAGE ROOMS (3.6)

◆   Division 2: Occupancies which present a moderate explosion hazard or a hazard from accelerated burning. (See code)

♦   Division 3: Occupancies which present a high fire or physical hazard.
(See code)

♦   Division 7: Occupancies which present health hazards. (See code)

♦   Separation of HPMs: Areas within a given storage room used for
differing classifications of HPMs -- toxics, acids, flammables,
oxidizers, etc. -- must also be separated.

♦   Article 79 of the UFC addresses requirements for flammable and
combustible liquids.

♦   Toxic Gas Monitoring: A continuous gas detection system must be
provided to detect the presence of hazardous (toxic or highly toxic)
gases. The detection system must initiate a local audible and visual
alarm as well as transmitting a signal to a constantly attended plant
Emergency Control Station (ECS).

♦   Activation of the alarm/monitoring system must automatically close
the shut-off valve on highly toxic or toxic gas supply lines.

♦   Smoke Detection: Supervised smoke detection must be provided in
rooms or areas where highly toxic compressed gases are stored
indoors.

♦   A dedicated telephone system which allows personnel to inform the
ECS of the exact nature and severity of an emergency condition is
recommended.

♦   Pyrophoric Materials: Requirements for pyrophoric materials storage
are covered in UFC Section 80.308.

HAZARDOUS MATERIAL DISPENSING, USE AND HANDLING (3.7)

♦   Article 80 Division IV addresses the dispensing, use and handling of hazardous materials.   In addition, UFC Article 79 Division VIII covers the use, dispensing and mixing of Class I, II or II-A flammable or combustible liquids both inside and outside of buildings.

♦   UFC 80.303 and 80.401 address special requirements for toxic and highly toxic gases:

   ▪   Piping and tubing utilized for highly toxic or toxic material shall have welded or brazed connections throughout unless an exhausted enclosure is provided.

   ▪   Treatment systems shall be utilized to handle the accidental release of gas.

   ▪   Several jurisdictions have enacted Toxic Gas Ordinances (TGO) which require that you comply with current UBC and UFC requirements for the storage, handling and treatment of gas releases, whether or not you modify or alter your facility.

HPM PIPING SYSTEMS (3.8)

♦   H-6 Occupancy General Requirements (UBC 911(f)) Piping: Piping shall be installed in accordance with nationally recognized standards. (An example of this would be the ANSI B31.3.)

♦   Where HPM supply gas is carried in pressurized piping, a fail-safe system shall shut off flow due to a rupture in the piping.

♦   Supply piping for HPM having a health hazard ranking of 3 or 4 shall have welded connections throughout unless an exhausted enclosure is provided.

♦   Each workstation utilizing HPM liquids shall <u>include</u>:
   ▪   Drainage piping to compatible system.
   ▪   Means to contain or direct spills to the drainage system.

4

Mechanical Heating, Ventilating and Air Conditioning Systems

4.1   **GENERAL:** The purpose of this chapter is to introduce and explain the requirements for mechanical HVAC systems for facilities using hazardous materials. The issues include:

1.   Exhaust Ventilation Systems.
2.   Emergency Shut-offs (*also refer to Chapter 6*).
3.   Air Handling System Isolation.
4.   Smoke and Fire Barriers.
5.   Exhaust Treatment and Isolation.
6.   Temperature Control (Required for HPMs).
7.   Smoke Control System.

Many hazardous occupancies, particularly as found in the semiconductor industry, contain cleanrooms or similar controlled environments. Cleanrooms utilize several types of air handling systems to create the desired environment:

-      Recirculating Air
-      Exhaust Ventilation
-      Make-up Air

Each of these systems must be properly coordinated with the other to ensure a safe occupancy is provided for people and the physical assets. The UBC and UFC stipulate requirements for ventilation rates, separation of air streams and the relationships between each system. The Uniform Mechanical Code provides additional technical requirements. The HVAC systems are a vital element in the code-mandated life safety package.

4.1.1   Recirculating Air Handling Systems:

The recirculating air streams in the cleanroom move significant rates of air through high efficiency filters (HEPA or ULPA) in a pattern designed to cause any particles generated in the space to take the shortest path out of the space. The high flow rates of cool air create special challenges for the fire suppression and smoke detection systems, as will be discussed later.

The recirculating air stream is considered an integral portion of the life safety system; therefore, in a hazardous occupancy cleanroom (such as a semiconductor wafer fab), the system is not to be automatically shut down by the smoke detection system. In lieu of automatic shutdown, a manual switch is to be provided outside the room, adjacent to the principal access door (UBC 905(b)). The return air system of one hazardous cleanroom area shall not connect to another cleanroom occupancy.

4.1.2   Exhaust Systems:

Rooms, areas or spaces in which explosive, corrosive, combustible, flammable or highly toxic dusts, mists, fumes, vapors or gases may be emitted, shall be mechanically ventilated (exhausted) as required by the Fire Code and the Mechanical Code. The design of the system shall be such that the emissions are confined to the area in which they are generated and shall be exhausted by a duct system to a safe location or treated by removing contaminants.

4.1.3   Make-up Air:

Make-up air is required to replenish air exhausted by the ventilation system. As a practical matter, the amount of make-up air required to replace the exhaust air and pressurize the cleanroom space may be significantly higher than the exhaust flow (typically 10% to 25% more than the exhaust flow rate, depending on the integrity of the construction).

4.2   RECIRCULATING AIR HANDLING SYSTEMS

There are few code requirements governing the recirculating air handling systems of a hazardous occupancy. These systems supply conditioned air to the occupied environment, then return that air to the heating/cooling unit for filtration and temperature control.

The primary issues of concern with regard to the recirculation systems is their role in the control of smoke and fire.

> ### UFC 51.105(d)2   (Fabrication Areas) Ventilation Requirements - Separate Systems
> "The return air system of one fab area shall not connect to another system within the building."

Discussion:

1.   Recirculating air systems for two or more fab areas must be independent of one another. A common air handling system utilized for a two-fab area <u>must</u> be a 100 percent outside-air system.

2.   The recirculating air system for an exit corridor shall be separated from the process areas of the fab.

> ### UFC 51.105(d)3 Ventilation Systems - Ventilation Controls
> "There shall be a manual control switch for the supply or recirculation air systems, or both, located outside of the fabrication area. The chief is authorized to require additional manual control switch locations."

A WORD OF CAUTION: The manual shutdown switch can be a source of potential hazard, or at least extreme nuisance, if a disgruntled or unauthorized employee decides to use it in a non-emergency situation. Consider placing the switch in the "emergency control station" where only qualified personnel have access. This will require concurrence of the code authorities.

4.3   EXHAUST VENTILATION SYSTEMS

4.3.1   Occupied Area Ventilation (Including Workstations):

The following code sections apply to requirements for ventilation of the occupied areas of hazardous facilities. Note that each section has somewhat different requirements. As with all code requirements, where there is an apparent contradiction, or variance in requirements, the more stringent criteria shall govern.

The most general criteria for ventilation of hazardous occupancies (UBC 905(b)) refers to the specific technical provisions of the Uniform Fire Code and Uniform Mechanical Code:

> **UBC 905(b)   (Requirements for Group H Occupancies) Light, Ventilation and Sanitation - Ventilation in Hazardous Locations**
>
> "Rooms, areas or spaces in which explosive, corrosive, combustible, flammable or highly toxic dusts, mists, fumes, vapors or gases are or may be emitted due to the processing, use, handling or storage of materials shall be mechanically ventilated as required by the Fire Code and the Mechanical Code.
>
> Emissions generated at workstations shall be confined to the area in which they are generated as specified in the Fire and Mechanical Codes.
>
> The location of supply and exhaust openings shall be in accordance with the Mechanical Code. Exhaust air contaminated by highly toxic material shall be treated in accordance with the Fire Code. *Article 80*

A <u>manual shutoff control</u> for ventilation equipment required by this subsection shall be provided outside the room adjacent to the principal access door to the room. The switch shall be of the break-glass type and shall be labeled 'Ventilation System Emergency Shutoff'."

The general criteria for semiconductor fabrication area ventilation is found in UBC 911(b)3. Note that the corollary requirements of the fire code are found in Article 51.

UBC 911(b)3 (Division 6 Occupancies) Fabrication Area - Ventilation

"Mechanical ventilation, <u>which may include recirculated air</u>, shall be provided throughout the fabrication area at the rate of not less than 1 cubic foot per minute, per square foot of floor area. The exhaust air duct system of one fabrication area shall not connect to another duct system outside that fabrication area within the building.

Ventilation systems shall comply with the Mechanical Code except that the automatic shutoffs need not be installed on air-moving equipment. However, smoke detectors shall be installed in the circulating airstream and shall initiate a signal at the emergency control station.

<u>Except for exhaust systems</u>, at least one manually-operated remote control switch that will shut down the fab area ventilation system shall be installed at an approved location outside the fabrication area."

Discussion:

1.   Although this section does not stipulate the amount of outside air which must be included in the total ventilation rate, the requirement of other codes (such as UFC as outlined below), govern the minimum exhaust rate. In reality, the total air flow and exhaust flow in a fab is typically far in excess of 1 CFM per square foot.

2.   The requirement for separation of exhaust systems for each "fabrication area" from other duct systems is based on several concerns:

    a.   The perimeter walls around the fab are fire-rated, and fab exhaust ducts generally contain corrosive materials, and, thus, cannot be fire dampened.

    b.   An exhaust fan failure could result in the migration of hazardous fumes from the fab to other areas, if the exhaust systems were interconnected.

3.   Automatic shutdown of the exhaust ventilation system upon detection of smoke is considered undesirable as the removal of the smoke can be aided by the system. Furthermore, the need to maintain negative pressure in the workstations to contain hazardous fumes is paramount to operator safety.

4.   A manual switch is required to allow the operators or fire department to shut down the ventilation system once the facility has been cleared of personnel.

> **UFC 51.105(d)1   Fabrication Areas - Ventilation Requirements**
> "Exhaust ventilation shall be provided to produce not less than one (1) cubic foot per minute, per square foot floor area, and shall be in accordance with Section 79.805(e)."

Discussion:

By definition, a "fabrication area" is an area within a Group H-6 occupancy in which there are processes involving HPMs and <u>may include</u> ancillary rooms such as dressing rooms, offices, etc., directly related to the fab area processes (UBC 407). We believe it is a good practice to provide general, or process, exhaust in <u>all</u> areas classified as H-6, at a rate of not less than 1 CFM/SF, or 6 air changes per hour, due to the potential for migration of hazardous fumes as a result of pressure differentials from one area to another.

UFC 51.106(c)1 (Workstations Within Fabrication Areas) Exhaust Ventilation - Design Criteria

"A ventilation system shall be provided to capture and exhaust fumes and vapors at workstations."

Discussion:

Generally, the manufacturer's requirements for process exhaust at each "workstation" or production tool far exceed the code requirements. Equipment densities in a typical wafer fab are such that exhaust ventilation generally far exceeds the 1 CFM/SF minimum, with 3 to 5 CFM/SF rates very common.

UFC 51.106(c)3 (Workstations within Fabrication Areas) Ventilation Power and Controls - Emergency Power

"The exhaust ventilation system shall have an emergency source of power . . . The emergency power is allowed to operate the exhaust system at not less than half fan speed (CFM) when it is demonstrated that the level of exhaust will maintain a safe atmosphere."

Discussion:

1.   *(Refer to Chapter 6 for a detailed explanation of emergency power requirements.)*

2.   The allowance for operation of the exhaust system under emergency power situations at half speed/CFM provides a means to reduce the size of engine/generator set. Practical and safe implementation of a half-flow emergency operation scenario may be more complicated and risky than can be justified. Some of the potential risks and complications include:

        a.   Difficulty in maintaining uniform air distribution in a large system; thus, tools or rooms nearest the fan may be adequately exhausted, while those remote from the fan may not be.

b.   Automatic controls and dampers in the exhaust system, which may be required to create a uniform exhaust flow condition, can be unreliable and costly.

c.   Some production equipment requires high negative pressure to operate effectively. The available pressure in the duct at half flow may not be adequate to ensure safe operation.

d.   The burden of proof to "demonstrate that the level of exhaust will maintain a safe atmosphere" is onerous.

e.   Changing from full-speed to half-speed operation -- or using one fan in lieu of two, then switching back to normal operation when power is restored -- exposes the facility to the risk of system failure on restart.

3.   While code requirements do not dictate that make-up air handling be provided with emergency power, practical operations and contamination control may dictate you do so. Upon a power failure, the exhaust system will continue to operate. Without emergency power for the make-up air fan, the fab pressure will become negative with respect to the surrounding atmosphere. Negative pressure in the fab may result in an excessive rate of particulate infiltration of the clean space, even if the power failure is of extremely short duration.

UMC 1105(a) Ventilation Systems and Product-Conveying Systems - Design

"A mechanical ventilation or exhaust system shall be installed to control, capture and remove emissions generated from product use or handling when required by the Building Code or Fire Code and when such emissions result in a hazard to life or property. The design of the system shall be such that the emissions are confined to the area in which they are generated by air currents, hoods or enclosures and shall be exhausted by a duct system to a safe location or treated by removing contaminants. Ducts conveying explosives or flammable vapors, fumes or dusts shall extend directly to the exterior of the building without entering other spaces.

Exhaust ducts shall not extend into or through ducts and plenums... Separate and distinct systems shall be provided for incompatible materials..."

Discussion:

1. The capture and removal of hazardous fumes is generally assured by the production tool or bench/workstation manufacturer. Design of the process exhaust system must take into account the negative pressure requirement at each point of connection and provide a means to effectively balance the system, given variations in the requirements.

2. Exhaust is generally discharged outside the building rather than treating to remove contaminants and recirculating, as treatment systems are costly and potentially unreliable.

3. The location and means of discharging air to the atmosphere must be carefully considered. High velocity (greater than 3,500 FPM) vertical discharge is generally provided to effect dilution of the potentially hazardous air stream with the ambient air. The relative location of exhaust discharge and fresh-air intakes to the building and adjacent buildings must be evaluated.

4. Separate duct/fan systems shall be provided for incompatible vapors and air streams. Pollution control ordinances may, in fact, have more jurisdiction over the design than safety considerations. Generally, fab exhaust systems are divided by the means of treatment, for example:

    a. Corrosive or general scrubbed exhaust.
    b. Solvent/Hydrocarbon exhaust, "abated" to remove the volatile organic compounds (VOCs) to the level acceptable to the local air pollution agency. The acceptable level of VOCs is generally expressed in terms of tons per year and will vary with the locale.

c.   Toxic exhaust, treated, incinerated or otherwise neutralized to preclude emissions from reaching the 1/2 IDLH (Immediately Dangerous to Life and Health) level. (*More on this later.*)

d.   General or "heat" exhaust, air streams without entrained hazardous materials.

5.   UMC 1105(a) requires that ducts which convey explosives or flammable vapors (toxics should be included in this group) shall extend directly to the exterior of the building, without entering other spaces. We believe it is <u>unacceptable</u> to connect gas cabinet exhaust to a workstation or any other exhaust ventilation system.

> **UMC 1105(b) Ventilation Systems and Product-Conveying Systems - Minimum Velocities and Circulation**
> "The velocity and circulation of air in work areas shall be such that contaminants are captured by an airstream at the area where the emissions are generated and conveyed into a product-conveying duct system. Mixtures within work areas where contaminants are generated shall be diluted below 25 percent of their lower explosive limit or lower flammability limit with air which does not contain other contaminants. The velocity of air within the duct shall be not less than set forth in Table No. 11-A...
> Systems conveying explosive or radioactive materials shall be pre-balanced through duct sizing. Other systems may be designed with balancing devices such as dampers. Dampers provided to balance air flow shall be provided with securely fixed minimum position blocking devices to prevent restricting flow below the required volume or velocity."

<u>Discussion:</u>

1.   Table 11-A sets no standards for duct velocity for systems conveying vapor, gases, smoke or fumes.

2.   The requirement for a self-balancing or fixed-adjustment exhaust system for ventilation of hazardous fumes is very important. Large systems in factories are frequently subject to variable duct pressure as the devices connected to the system are changed, or operate through a cycle wherein flow rates change. Fluctuations in the exhaust duct pressure will result in fluctuations in the exhaust flow at the process tool, gas cabinet, hood or other device. If the system pressure fluctuations are severe, the flow may not be reliably maintained. We believe careful consideration should be given to the use of reliable pressure independent flow controllers. Of particular concern is the need to assure reliable flow at gas cabinets with toxic, pyrophoric and otherwise hazardous gases.

UMC 1106 Ventilation Systems and Product-Conveying Systems - Hoods and Enclosures

"Hoods and enclosures shall be used when contaminants originate in a concentrated area. The design of the hood or enclosure shall be such that air currents created by the exhaust systems will capture the contaminants and transport them directly to the exhaust duct. The volume of air shall be sufficient to dilute explosive or flammable vapors, fumes or dusts as set forth in Section 1105(b). . . ."

Discussion:

It is obvious that the preferred means of removing hazardous fumes is with a local hood at the point of use. Manufacturing production tools are generally provided with an abundance of exhaust to capture the vapors or gases generated by the process. Wet hoods or benches provided in etching, cleaning and stripping processes are generally exhausted at a rate of 100 to 150 CFM per square foot of hood face area, depending on the process and the hazardous materials handled. The users' safety department generally has requirements for exhaust flow rates which exceed those stipulated by the codes. The challenge in today's economic environment is to reduce the rate of exhaust, in a safe manner, such that the cost of constructing and operating a manufacturing facility is reduced.

Conclusion:

UBC 911(b)3 (which is the general reference for fabrication area ventilation) is the <u>least</u> stringent of the code requirements. It requires one (1) CFM per square foot of floor area and allows some of this (an unspecified amount) to be recirculated air.

UBC 905(b) is more strict than 911 in that it refers to the UFC and UMC where hazardous materials are used and vapors may exist. There is a potential for hazardous vapors to occur or be diffused throughout areas in which such materials are not specifically used. We believe it is good design practice to assume potentially harmful vapors may exist in all areas of an "H occupancy," thus, the stipulated exhaust and make-up air flow should be provided throughout.

UFC 51.105(d) is the <u>most</u> stringent ventilation specification. It requires <u>exhaust</u> ventilation at the rate of one (1) CFM per square foot of floor area (as a minimum) and further requires compliance with UFC 79.805(e).

These exhaust/ventilation requirements apply not only to the fabrication work zones (clean bays), but also to equipment rooms, utility zones and any other rooms within the "H" occupancy. In fabrication area work zones, these requirements are usually met by process exhaust rates. If the process exhaust is not at least 1 CFM/SF, area exhaust must be provided in addition to equipment exhaust to meet the minimum requirements. Such area exhaust should be taken from a point at or near the floor, as the majority of hazardous vapors are heavier than air.

4.3.2   HPM Storage Area Ventilation:

Chemical storage areas consist of inside liquid storage rooms, liquid storage rooms and H-1, H-2, H-3 or H-7 HPM storage rooms. (*Refer to Chapters 2 and 3 for a general discussion of chemical storage room construction, location and exiting requirements.*)

When the room is used for storage of Type I, II or III-A flammable liquids in <u>closed</u> containers, pressurized at <u>less</u> than 15 psig, it will generally be classified as Group H-3, if the quantities exceed those outlined in UBC Table 9-A. The room may be classified as a "control area" if the quantities are less than the "exempt" amounts. Exhaust ventilation (and adequate make-up air) will be required as outlined below.

When the room houses Type I, II or III-A flammable liquids which are used in <u>open</u> containers or closed containers at <u>more</u> than 15 psig, then the room is generally classified as Group H-2 if the quantities exceed those outlined in UBC Table 9-A. The room may be classified as a "control area" if the quantities stored are less than the exempt amounts. The ventilation rates for H-1, H-2, H-3 and H-7 storage are as follows:

UBC 905(b) (Requirements for Group H Occupancies) Light, Ventilation and Sanitation - Ventilation in Hazardous Locations

"Rooms, areas or spaces in which explosive, corrosive, combustible, flammable or highly toxic dusts, mists, fumes, vapors or gases are, or may be, emitted due to the processing, use, handling or storage of materials shall be mechanically ventilated as required by the Fire Code and the Mechanical Code."

UBC 911(e)4 (Division 6 Occupancies) Storage of HPM - Ventilation

"Mechanical exhaust ventilation shall be provided in storage rooms at the rate of not less than one (1) cubic feet per minute per square foot of floor area or six (6) air changes per hour, whichever is greater, for all categories of material."

UFC 51.110(b)5   Storage of Hazardous Production Materials Within Buildings - Ventilation Requirements

"Ventilation shall be provided in accordance with Section 51.105(d)1." *(See reference above.)*

UFC 79.203(g) Flammable and Combustible Liquids, Liquid Storage Rooms - Ventilation

"Liquid storage rooms shall be ventilated in accordance with Section 80.301."

UFC 80.301(m)1-7 Hazardous Materials, Storage - Ventilation

"Unless exempted or otherwise provided for in Sections 80.302 through 80.315, indoor storage areas and storage buildings shall be provided with mechanical exhaust ventilation.

> **Exception:** Where natural ventilation can be shown to be acceptable for the materials as stored.

Exhaust ventilation systems shall comply with the following:

1. Installation shall be in accordance with the provisions of the Mechanical Code.

2. Mechanical ventilation shall be at a rate of not less than 1 cubic foot per minute per square foot of floor area over the storage area.

3. Systems shall operate continuously. Alternate designs may be approved by the chief.

4. A manual shutoff control shall be provided outside the room adjacent to the access door into the room or in a location approved by the chief. The switch shall be of the break-glass type and shall be labeled 'Ventilation System Emergency Shutoff.'

5. Exhaust ventilation shall be arranged to consider the density of the potential fumes or vapors released. For fumes or vapors that are heavier than air, exhaust shall be taken from a point within 12 inches of the floor.

6.   The location of both the exhaust and inlet air openings shall be arranged to <u>provide air movement across all portions of the floor</u> or room to prevent the accumulation of vapors.

7.   Exhaust ventilation <u>shall not be recirculated within the room or building if the materials stored are capable of emitting hazardous vapors.</u> "

Discussion:

1.   We believe the <u>minimum</u> exhaust flow rate should be one (1) CFM/SF or not less than six (6) air changes per hour if the room height exceeds 10 feet.

2.   The exhaust ventilation system should be connected to <u>emergency</u> power supply to ensure continuous operation.

3.   We prefer the manual shut-off control switch be located at the emergency control station, with the general ventilation control switch.

Storage rooms containing more than the exempt amounts of toxic or highly toxic solids or liquids are classified H-7. The requirement for ventilation is given in UFC Article 80.

UFC80.312(a)7 (Highly Toxic Solids and Liquids) Indoor Storage - Exhaust Scrubber

"Exhaust scrubber or other systems for the processing of highly toxic liquid vapors shall be provided for storage areas where a spill or other accidental release of such liquids can be expected to release highly toxic vapors . . ."

Discussion:

(*Requirements for exhaust treatment and scrubbing are outlined in Section 4.4.*) Local air quality authorities should also be consulted.

4.3.3 Gas Cabinet and Enclosure Ventilation:

The use of gas cabinets for the storage and dispensing of health and physical hazard gases is recommended *as discussed in Chapter 3*. When the amount of toxic or highly toxic gas exceeds that listed as "exempt" in UFC 80.303(a)2, the occupancy is classified H-7. When the amount of physical hazard gas (flammable or pyrophoric) exceeds the "exempt" amount, the occupancy is classified H-2. Code provisions for the ventilation of such enclosures are discussed below.

UFC 51.107(b)3 (Storage and Dispensing of HPM Within Fabrication Areas) Special Requirements for HPM Gases - Ventilation

"Gas cabinets shall be provided with ventilation. When a gas cabinet contains highly toxic or toxic gases, the average velocity of ventilation at the face of access ports shall be not less than 200 feet per minute (fpm) with a minimum of 150 fpm at any point of the access port. .... The gas cabinet ventilation system shall be in accordance with Sections 51.106(c)2 and 3 and is allowed to connect to a workstation ventilation system."

<u>Discussion:</u>

1. Although this section only specifically refers to "toxic" gases to be ventilated at 200 fpm, general practice is to enclose all gases except inerts in gas cabinets and apply the same ventilation rate. The theory is that most gases, corrosives, toxics, pyrophorics, etc., are indeed hazardous to occupants and need to be contained via the ventilation system.

2. The exhaust system must provide adequate face velocity not only at the access port (to the valve), but through the door of a multiple cylinder cabinet when it is opened to replace a bottle. The design of the ventilation system may need to provide a two position (flow) operation interlocked with the door to increase the flow through the cabinet when the door is opened.

3.   While the UFC "allows" connection of the gas cabinets to the general fabrication area exhaust system, we urge caution in this area, as there is the potential for cross contamination, fire or explosion if a catastrophic release of gas were to occur. Additionally, the potential for chemically reactive mixtures must be carefully evaluated.

4.   Pyrophoric Gas Storage: Experience has shown that exhaust for gas cabinets with pyrophoric gases should be treated specially. The exhaust of silane gas cabinets is particularly critical, in that silane has been known to create a protective "bubble" and then detonate, rather than burn, when the exhaust ventilation rate in the area of a leak is inadequate. It has been shown that air velocity in the region of a silane leak of at least 200 feet per minute will ensure adequate mixing with air. This will prevent the formation of a protective bubble and, thus, the silane will burn, rather than detonate. Accordingly, we recommend silane and pyrophoric gas cabinets be exhausted at 600 to 700 CFM for two-bottle cabinets and 900 to 1,000 CFM for three-bottle cabinets, at all times.

Care must be taken to ensure the flow rate through the cabinet will not fall below these values when the access port is closed. Most cabinets have different exhaust flow characteristics dependent upon whether the access port is open or closed.

UFC 80.303(a)6A (Toxic and Highly Toxic Compressed Gases) Indoor Storage - Ventilation and Storage Arrangement

"Storage of cylinders shall be within ventilated gas cabinets, exhausted enclosures or within a ventilated separate gas storage room. ...If gas cabinets are provided, the room or area in which they are located shall have independent exhaust ventilation."

UFC 80.303(a)6B(i-iii) (Indoor Storage) Ventilation and Storage Arrangement - Gas Cabinets
"When gas cabinets are provided, they shall be:

(i)   operated at negative pressure in relation to the surrounding area,

(ii)   provided with self-closing limited access ports . . . The average velocity of ventilation at the face of access ports or windows shall be not less than 200 feet per minute with a minimum of 150 feet per minute at any point,

(iii)   connected to an exhaust system, . . . "

UFC 80.402(b)3G(i-iv, viii) (Closed Systems) Special Requirements for Highly Toxic and Toxic Compressed Gases

(i)   Ventilation and Storage Arrangement. "Compressed gas cylinders in use shall be within ventilated gas cabinets, laboratory fume hoods, exhausted enclosures or separate gas storage rooms . . ."

(ii)   Gas Cabinets and Exhaust Enclosures. "When gas cabinets or exhausted enclosures are provided, they shall be in accordance with Section 80.303(a)6B. Gas cabinets and exhausted enclosures shall be internally-sprinklered. "

(iii)   Separate Gas Storage Rooms. "When separate gas storage rooms are provided, they shall be in accordance with Section 80.303(a)6C. "

(iv)   Treatment Systems. "Treatment systems shall be provided in accordance with Section 80.303(a)6D. " (*Refer to Section 4.6 for discussion of requirements.*)

(viii)   Process Equipment. "Effluent from process equipment containing highly toxic or toxic gases which could be discharged to the atmosphere shall be processed through an exhaust scrubber or other processing system . . ."

Discussion:

The general industry practice for treatment of the effluent from each process tool with toxic gases is to provide a point-of-use (POU) dedicated specialty scrubber. Such scrubbers are generally very high in removal efficiency, but low in flow capability. The treated discharge from such POU scrubbers is then generally directed to the "house" scrubber, where additional treatment of the effluent and dilution render it safe for discharge to the atmosphere. (*Refer to* **Figures 4.1 and 4.2** *for examples of commercially available equipment.*)

UFC 80.402(c)8A (Exterior Dispensing and Use) Special Requirements for Highly Toxic and Toxic Compressed Gases - Ventilation and Storage Arrangement
"When cylinders or portable containers are used out-of-doors, gas cabinets or a locally exhausted enclosure shall be provided."

UFC 80.402(c)8B (Exterior Dispensing and Use) Special Requirements for Highly Toxic and Toxic Compressed Gases - Gas Cabinets
"When gas cabinets are provided, the installation shall be in accordance with Section 80.303(a)6B."

UFC 80.402(c)8C (Exterior Dispensing and Use) Special Requirements for Highly Toxic and Toxic Compressed Gases - Treatment Systems
"Treatment systems shall be provided in accordance with 80.303(a)6D."

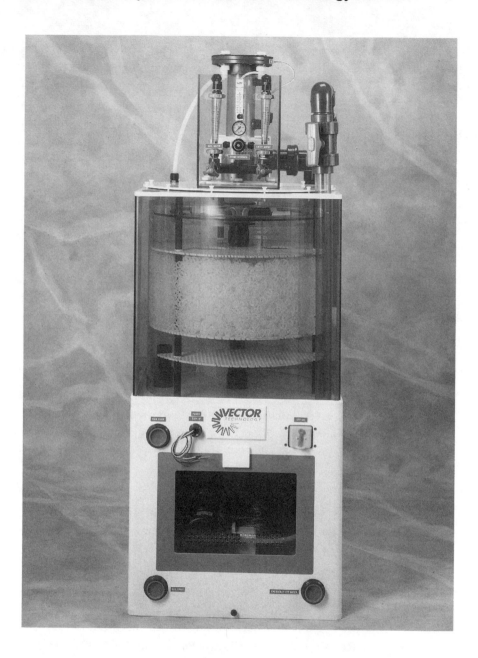

Figure 4.1   POINT-OF-USE SCRUBBER for Removing HCl or Other Reactive Gases from Process Tool Exhaust.   (Photo courtesy of Vector Technology)

Figure 4.2 FOUR-STACK SCRUBBING SYSTEM for Four-Chamber "Cluster Tool" Point-of-Use Toxic Gas Mitigation. (Photo courtesy of Ecoloteck, Inc.)

UFC 80.303(a)6C(i-ii)   (Indoor Storage) Ventilation and Storage Arrangement - Separate Gas Storage Rooms

"When separate gas storage rooms are provided, they shall be designed to:

(i)   Operate at a negative pressure in relation to the surrounding area, and

(ii)   Direct the exhaust ventilation to an exhaust system."

UFC 80.303(a)6D   (Indoor Storage) Ventilation and Storage Arrangement - Treatment Systems

(Note: Treatment systems are discussed in detail in Section 4.6.2.)

UFC 80.303(c)3A-B (Indoor Storage) Special Provisions - Gas Cabinets for Leaking Cylinders

"At least one gas cabinet or exhausted enclosure shall be provided for the handling of leaking cylinders . . . The gas cabinet or exhausted enclosure shall be connected to an exhaust system.   See Section 80.303(a)6D."   (Treatment system)

UFC 80.303(c)4A-B (Indoor Storage) Special Provisions - Local Exhaust for Leaking Portable Tanks

"A means of local exhaust shall be provided to capture leaks from portable tanks.   Portable ducts or collection systems designed to be applied to the site of a leak in a valve or fitting on the tank are acceptable.   The local exhaust system shall be connected to a treatment system as specified in Section 80.303(a)6D.   The local exhaust system shall be provided:

A   Within or immediately adjacent to exterior storage areas, or

B   Within separate gas storage rooms used for portable or stationary tanks."

4.3.4   Chemical Use, Dispensing and Mixing Room Ventilation:

For the purpose of our discussions, it is important to understand the "official" definition of Use, Dispensing and Mixing:

> "USE (Material) is the placing in action or making available for service by opening or connecting anything utilized for confinement of materials whether a solid, liquid or gas." (UFC 9.123)

> "DISPENSING is the pouring or transferring of a material from a container, tank or similar vessel whereby vapors, dusts, fumes, mists or gases may be liberated to the atmosphere." (UFC 9.106)

Basically, any activity involving HPMs, other than passive storage in unopened original containers, can be considered Use, Dispensing or Mixing, and is, therefore, governed by the 80.400 series of requirements in Article 80.

> **UFC 79.805(e)   Use, Dispensing and Mixing Rooms - Ventilation**
> "Continuous mechanical ventilation shall be provided at a rate of not less than 1 cubic foot per minute, per square foot of floor area over the design area. For ventilation system design, see the Building and Mechanical Codes."

Discussion:

It is presumed that the term "ventilation," as used here, refers to exhaust air and not recirculated air (refer to UFC 80.301(m)7).

> **UFC 79.1304(a) (Industrial Plants) Ventilation - Open Systems and Processes**
> "Building or rooms in which Class I, II or III-A liquids are used or stored in open systems and processes shall be provided with ventilation in accordance with this section ..."

UFC 79.1304(b) (Industrial Plants) Ventilation - Design

"Design of ventilating systems shall consider the relatively high specific gravity of the vapors... Mechanical systems for removing flammable and combustible liquid vapors shall be designed and installed in accordance with the Mechanical Code."

UFC 80.402(b)2C (Indoor Dispensing and Use) Open Systems - Ventilation

"When gases, liquids or solids having a hazard ranking of 3 or 4 in accordance with UFC Standard No. 79-3 are dispensed or used, mechanical exhaust ventilation shall be provided to capture fumes, mists or vapors at the point of generation.

> **Exception:** Gases, liquids or solids which can be demonstrated not to create harmful fumes, mists or vapors."

UFC 80.402(b)3C Indoor Dispensing and Use - Closed Systems - Ventilation

"If closed systems are designed to be opened as part of normal operations, ventilation shall be provided in accordance with the provisions of Section 80.402(b)2C."

4.3.5 Corridor Ventilation:

Code sections which apply to service and exit corridors include the following:

"UBC 911(d) Division 6 Occupancies - Service Corridors

"Service corridors shall be classified as Group H, Division 6 occupancies...

Service corridors shall be mechanically ventilated as required by Section 911(b)3 or at not less than six (6) air changes per hour, whichever is greater..."

Discussion:

1.   Service corridors are defined as those passageways used for transporting HPMs and not for exiting (UBC 420).   (*Refer to complete discussion in Chapter 2 of this book.*)

2.   The requirements of UBC 911(b)3 must be met in a service corridor. This requirement is one (1) CFM, per square foot of floor area.

3.   By virtue of classification as an H-6 occupancy,   UFC 51.105(d) requires that the "fabrication area" be provided with one (1) CFM per square foot exhaust ventilation.   Recirculation is not to substitute for exhaust, although recirculation is frequently provided, as the service corridors may be Class 100 to Class 1,000 clean spaces.   The UFC definition (9.108) of fabrication area includes ancillary rooms in which there are "processes involving hazardous production materials" (such as HPM delivery).

> **UBC 911(c) Division 6 Occupancies - Exit Corridors**
> "Exit corridors shall comply with Section 3305...shall not be used for transporting hazardous production materials..."

Discussion:

1.   The general H-6 requirement for "ventilation" (UBC 911(b)3) at a rate of not less than 1 CFM/SF could be interpreted to allow recirculation air, in lieu of exhaust or outside air, for an exit corridor, as there are no specific code citations for the ventilation of exit corridors.   We believe it is unwise not to exhaust the exit corridor in the same manner as other H-6 spaces which have HPMs.   The rationale for this is that the exit corridors are typically operated at a lower air pressure than the adjacent clean fab spaces; thus, when doors are opened, the fab air may exfiltrate into the exit corridor, potentially carrying hazardous fumes.

2.   The UFC requirement for no less than 1 CFM/SF exhaust ventilation (UFC 51.105(d)1) for all H-6 fabrication areas should be applied when the exit corridor is within the H-6 occupancy.

3.   The recirculating air handling system of the exit corridor must be separated from the recirculating air system of other areas of the fab, in accordance with UFC 51.105(d)2.

> **UBC 911(f)2A-B,D (Division 6 Occupancies) Piping and Tubing - Installations in Exit Corridors and above Other Occupancies**
>
> "Hazardous production materials shall not be located within exit corridors or above areas not classified as Group H, Division 6 Occupancies except as permitted by this subsection. Hazardous production material piping and tubing may be installed within the space defined by the walls of exit corridors and the floor or roof above or in concealed spaces above other occupancies under the following conditions:"
>
> A   Automatic Sprinklers.
>
> B   "Ventilation of not less than 6 air changes per hour shall be provided. The space shall not be used to convey air from any other area."
>
> D   "All HPM supply piping ... shall be separated from the exit corridor . . . by construction ... of not less than one hour..."

Discussion:

This reference for exit corridors requires (exhaust ventilation) at a rate of six (6) air changes per hour for the enclosed space, above the ceiling of the exit corridor where HPM chemicals and/or gases are being transported in approved piping.

4.4   MAKE-UP AIR SYSTEMS

Make-up air systems serve several very important functions in a hazardous occupancy.

First and foremost, the system is used to replace the air exhausted from the occupancy. Without an adequate, powered make-up air system, the space would operate at negative pressure and the ability to exhaust the required flow rate would be jeopardized.

Second, the make-up air system is generally used to condition the outside air prior to introduction into the occupied environment. The extent of the required conditioning is dependent on the environmental control specifications of the space. In semiconductor manufacturing or similar spaces where stringent control of temperature and relative humidity is dictated, the make-up air system provides heating, cooling, humidification, dehumidification and filtration of the outside air.

> **UMC 1105(c) Ventilation Systems and Product-Conveying Systems - Make-up Air**
> "Make-up air shall be provided to replenish air exhausted by the ventilation system. Make-up air intakes shall be located so as to avoid recirculation of contaminated air within enclosures."

Discussion:

1.   The make-up air system is crucial to not only life safety in the fab, but control of the process environment in terms of particulates, temperature and humidity.

2.   The location of fresh-air intakes relative to the points of exhaust and other contaminants, such as vehicle exhaust, emergency generators, etc., should be carefully considered. A computer-aided model of the facility which accounts for building geometry, winds and other influences may be beneficial.

Outside-air quantities in excess of exhaust quantities must be supplied to work zones of cleanrooms for pressurization to keep particles out and to allow environmental conditions to be controlled to close tolerances. UMC 1105(c) requires make-up air to replace exhaust air. As a practical matter, the amount of make-up air required to replace the exhaust air and pressurize the space may be significantly higher than the exhaust flow. Depending on the type of construction (walls and ceiling) and the care taken to seal around penetrations of the envelope for the piping, conduit and ductwork, the exfiltration of clean air from the space may exceed the amount of exhaust air. The cost of providing make-up air in terms of capital (for equipment) and expense for energy is significant.

Based on experience with a variety of wafer fab facilities, the following total process exhaust and ventilation rates are offered for reference:

FACILITY	FAB (H-6) AREA (SQ. FT.)	CFM EXHAUST	CFM OUTSIDE AIR	CFM/SF EXHAUST	CFM/SF OUTSIDE AIR
Fab 1	28,400	71,000	90,000	2.5	3.2
Fab 2	28,350	80,000	90,000	2.8	3.1
Fab 3 (Partial)	7,500	25,000	33,000	3.3	4.4
Fab 4	36,480	90,000	125,000	2.5	3.4

4.5   EMERGENCY VENTILATION AND OPERATION

4.5.1   Emergency power is required to operate the exhaust ventilation systems in H-6 occupancies. Emergency power must be on-line within ten seconds of failure of the primary power source. The start-up of standby power can be delayed for up to 60 seconds after a power failure.

UFC 51.106(c)3 Workstations within Fabrication Area - Ventilation Power and Controls - Emergency Power

"The exhaust ventilation system shall have an emergency source of power. The emergency power shall be designed and installed in accordance with the Electrical Code. The emergency power is allowed to operate the exhaust system at not less than one-half fan speed when it is demonstrated that the level of exhaust will maintain a safe atmosphere."

Not commercial →

Discussion:

Emergency power is required so that manufacturing processes which utilize hazardous materials will continue to be exhausted upon loss of normal power. If exhaust is being handled by scrubbers or a solvent removal system, emergency exhaust fans should be considered. These fans may be permitted to bypass the scrubbers or abatement systems, eliminating substantial pressure drop, reducing fan horsepower required and, ultimately, reducing required emergency generator capacity. This type of exhaust bypass system needs the approval of the local air pollution control agency.

We believe that fab area pressurization should be maintained to control cleanliness. Therefore, it is recommended that each facility consider providing outside make-up air in excess of exhaust air quantities on emergency power. An emergency power air handler for outside air can be utilized.   The emergency power make-up air could be furnished without coils and HEPA filters to reduce fan power requirements. (Remember that the central cooling and heating plant will most likely shut down as a result of power failure). Pressure drop and fan horsepower of this air handler can be reduced with corresponding reduction in required emergency generator capacity.

The allowance for operating the exhaust system at "not less than half speed" is presumed to mean that the system will operate at not less than half of the code-required flow rate.   The stipulation "when it is demonstrated that the level of exhaust will maintain a safe atmosphere" places the burden of proof on the designer and owner. We have generally found that half the normal air flow still exceeds the required 1 CFM/SF exhaust; however, the need for proof of "a safe atmosphere" will depend on the building official. The use of hoods and chemical cabinets will generally help ensure the safety of the atmosphere; however, this may not be sufficient proof. (*This issue is further discussed in Section 4.3 above.*)

The facilities utilized to store compressed toxic gases must be in accordance with the provisions of UFC 80.303(a) and (c).

> ### UFC 80.303(a)1   Toxic and Highly Toxic Compressed Gases - Indoor Storage
> "Indoor storage of toxic and highly toxic compressed gases shall be in accordance with this subsection, and Sections 80.301 and 80.303(c)."
>
> ### UFC 80.303(a)7A-D   (Toxic and Highly Toxic Compressed Gases) Indoor Storage - Emergency Power
> "Emergency power shall be provided in lieu of standby power for:
>
> A   Exhaust ventilation, including the power supply for treatment systems,
>
> B   Gas-detection systems,
>
> C   Emergency alarm systems, and
>
> D   Temperature-control systems."

It is prudent, if not mandatory, to provide emergency power, or preferably an uninterruptible power supply (UPS) for all controls which operate the life safety related equipment.

4.5.2   Standby power is required to be provided for areas where hazardous materials are stored. Emergency power is required (in lieu of standby power) whenever highly toxic materials are used or dispensed. The distinction between standby power and emergency power is discussed in detail in Chapter 6 of this book; however, the basic difference is that the start-up of standby power can be delayed for up to 60 seconds after a power failure. This time delay allows non-simultaneous starting of motor loads which reduces the inrush current loading and required size of the generator.

UFC 80.301(s) Storage - Standby Power

"When mechanical ventilation, treatment systems, temperature control, alarm, detection or other electrically-operated systems are required, such systems shall be connected to a secondary source of power to automatically supply electrical power in the event of loss of power from the primary source. See the Electrical Code."

UFC 80.401(l) Dispensing, Use and Handling - Standby and Emergency Power

"When mechanical ventilation, treatment systems, temperature control, manual alarm, detection or other electrically operated systems are required by other provisions of this division, such systems shall be connected to a standby source of power to automatically supply electrical power in the event of loss of power from the primary source. See the Electrical Code.

When highly toxic compressed gases or highly toxic, highly volatile liquids are used or dispensed, <u>emergency power</u> shall be provided in lieu of standby power on all required systems. See the Electrical Code."

It is our opinion and recommendation that the exhaust and make-up air systems for the HPM storage areas be provided with emergency power. The code allows these facilities to be provided with "standby power" because they are generally not occupied on a full-time basis.

4.5.3 Emergency-Power-Off Scenario:

There are various references in both the UBC and UFC which require controls to shut down the supply (primary and secondary) and exhaust air handling systems. (*These requirements are discussed in Chapter 6, ELECTRICAL.*)

4.6   EXHAUST TREATMENT AND SEPARATION OF REACTIVE MATERIALS

4.6.1   Separation:

Separation of incompatible air streams such as acid and solvent exhaust is required by various regulations.

UFC 51.106(c)2B   (Exhaust Ventilation) Duct Systems - Reactives

"Two or more operations shall not be connected to the same exhaust system when either one, or the combination of the substances, removed may constitute a fire, explosion or chemical reaction hazard within the duct system."

UMC 1105(a) Ventilation Systems and Product-Conveying Systems - Design

"...Ducts conveying explosives or flammable vapors, fumes or dusts shall extend directly to the exterior of the building without entering other spaces. Exhaust ducts shall not extend into or through ducts and plenums.

> **Exception:** Ducts conveying vapor or fumes having flammable constituents less than 25 percent of their lower flammability limit (LFL) may pass through other spaces.

Separate and distinct systems shall be provided for incompatible materials..."

While it may not be a code requirement (based on strict interpretation) that acid and solvent exhaust systems be isolated (because the reactives are usually too dilute to constitute a hazard), good engineering practice and pollution abatement regulations generally require separation of solvent and corrosive exhaust.   Corrosive exhaust ducts are generally constructed of fiberglass reinforced plastic (FRP), which is not suitable for solvent vapors. Additionally, most air pollution control jurisdictions require fume scrubbers to remove corrosive materials from the effluent air stream and solvent extraction processes to remove hydrocarbons from the exhaust. Solvent extraction and acid fume scrubbers are not compatible; therefore, separation of the solvent and acid systems is dictated.

When feasible, a heat exhaust system should be provided to ventilate spaces or equipment which do not contain corrosive or hydrocarbon vapors. This technique can reduce the installed costs (scrubbers and solvent abatement equipment are very expensive) and the operating cost (abatement devices generally have very high pressure losses and corresponding high fan power requirements).

4.6.2   Treatment:

The Uniform Fire Code (UFC), Section 80.303(a)6D requires a "treatment system" to process the exhaust discharged from gas cabinets, exhausted enclosures or separate storage rooms. The requirement is to reduce the discharge concentration of the gas released from a gas cylinder failure to not more than one-half the concentration considered Immediately Dangerous to Life and Health (1/2 IDLH) at the point of discharge. Many of the exhaust systems currently in place for gas storage facilities are not "treated;" therefore, this represents a significant new requirement and expense. This requirement is addressed in the Toxic Gas Ordinances (TGO) of several jurisdictions. The TGO, essentially, requires the retrofit of "treatment" systems, whether or not the facility is renovated or otherwise modified.

> **UFC 80.303(a)6D(i,ii)   (Indoor Storage) Ventilation and Storage Arrangement - Treatment Systems**
> i General. "Treatment systems shall be utilized to handle the accidental release of gas. Treatment systems shall be utilized to process all exhaust ventilation to be discharged from gas cabinets, exhausted enclosures or separate storage rooms."

ii   Design.   Treatment systems shall be capable of diluting, adsorbing, absorbing, containing, neutralizing, burning or otherwise processing the entire contents of the largest single tank or cylinder of gas stored or used . . ."

UFC 80.303(a)6D(iii)   (Ventilation and Storage Arrangement) Treatment Systems - Performance

"Treatment systems shall be designed to reduce the maximum allowable discharge concentration of the gas to one-half IDLH at the point of discharge to the atmosphere.   When more than one gas is emitted to the treatment system, the treatment system shall be designed to handle the worst case release based on the release rate, the quantity and the IDLH for all the gases stored or used."

[handwritten margin note: IMMEDIATELY DANGEROUS TO LIFE & HEALTH]

UFC 80.303(a)6D(iv)   (Ventilation and Storage Arrangement) Treatment Systems - Sizing

"Treatment systems shall be sized to process the maximum worst case release of gas based on the maximum flow rate of release from the largest cylinder or tank utilized.   The entire contents of tanks and cylinders shall be considered."

TABLE NO. 80.303-B
RATE OF RELEASE FOR CYLINDERS AND PORTABLE TANKS

Container	Nonliquefied (Minutes)	Liquefied (Minutes)
Cylinders	5	30
Portable Tanks	40	240

Treatment system methodologies are not discussed in the UFC except to say that "such exhaust systems shall be capable of diluting, adsorbing, absorbing, neutralizing, burning or otherwise processing the entire contents of the largest single tank or cylinder of gas stored."

Dilution is rarely effective as the sole means of treatment of an unrestricted cylinder release, due to the great amount of dilution required. Using a 90-pound cylinder of chlorine as a typical example, and calculating a release rate under standard temperature and pressure conditions, our governing criteria are:

1.   30 minute release time, from Table 80.303-B.
2.   90 pounds of chlorine released (complete discharge).
3.   Gas density of 5.4 cubic feet/pound.
4.   IDLH level of 25 PPM.

The release rate is 90 lb./30 min. = 3 lb./min., times 5.4 ft^3/lb = 16.2 CFM of chlorine. To dilute the concentration to one-half the IDLH level, or 12.5 PPM (one part in 80,000), 16.2 times 80,000 or 1,296,000 CFM of exhaust is required, using dilution as the sole treatment!

Dilution can still play a role in the treatment process, especially in larger systems. For a large gas storage room that has a 25,000 CFM exhaust ventilation rate, the permissible chlorine discharge rate in the above example would be 25,000 CFM x 12.5 PPM = 0.3125 CFM chlorine. The upstream treatment system must then treat to at least this discharge rate, requiring an efficiency of [(3 - 0.3125)/3 x 100] percent, or 89.6 percent. It is important to note that the gas concentration is measured at the point of discharge to atmosphere. If the gas cabinets were individually ventilated, the individual exhaust flow rates would be smaller, and so would the permissible gas discharge rates. The treatment systems used, then, would have to be correspondingly more efficient.

Note that concentration levels are determined by volume, not mass. Most HPM gases are heavier than air; thus, a gas concentration by weight is more dilute than an equal concentration by volume.

If a restricted flow orifice (RFO) is placed in the cylinder, the rate of release during a catastrophic event is reduced substantially. As an example, if a 0.01-inch orifice is used with chlorine at an initial cylinder pressure of 100 psia, the flow rate will be 0.36 CFM, in lieu of the 16.2 CFM stipulated by UFC 80.303(b). This release rate can be diluted by an exhaust stream of 29,000 CFM, in lieu of 1,296,000 CFM discussed above. This is now within the realm of reality. An RFO with orifice size of 0.006-inch would reduce the flow to 0.22 CFM, requiring an exhaust flow rate of only 17,600 CFM! These flow rates are calculated without any scrubbing or other means of chemical removal from the air stream, which if utilized, would substantially reduce the required exhaust flow rate.

The traditional method of removing acids and fumes from air streams has involved wet scrubbing, using water or a water-based chemical solution. In this approach, a contaminated airstream is thoroughly mixed with water or other chemical solution to absorb and/or neutralize the contaminant and subsequently remove it from the airstream. The water solution is recirculated, with appropriate controls to maintain solution pH, total dissolved solids (TDS) level, etc. This traditional method has worked well for airstreams containing low concentrations of contaminants, on the order of 10 PPM, with standard equipment ranging in size from 1,500 to 100,000 CFM.

Since a catastrophic release may have an exhaust contaminant concentration that is many times higher than traditional scrubbed exhaust airstreams (1600 PPM for the unrestricted chlorine cylinder above, at 25,000 CFM) some enhancement of conventional equipment is necessary to increase scrubber efficiencies. Two differing approaches have been used in upgrading the wet scrubber process to handle gas cabinet catastrophic releases.

One approach is to scrub the exhaust from all the gas cabinets, with an increased packing depth and, usually, a chemical feed system to more quickly neutralize and remove the potential contaminants encountered. This approach is relatively reliable (since the system operates whether contaminants are present or not), and low in initial cost (approximately $4 to $5/CFM), but requires a large amount of room for equipment. Maintenance cost is moderate due to chemical use and the need to keep the system clean. Energy cost is relatively low.

Treatment of toxic metal hydrides and/or pyrophorics present special problems in that the contaminants are generally not removed using standard chemistries for a scrubbing process. Generally, oxidation-reduction systems are employed with scrubbers using potassium permanganate ($KMnO_4$) or, less frequently, sodium hypochlorite (NaOCl) as an oxidizing agent in combination with sodium hydroxide feed systems. Both oxidizers are somewhat problematic due to the permanganate's low solubility (4 percent maximum) and the hypochlorite's high volatility (sodium hypochlorite is the active ingredient in bleach). Equipment unit costs are considerably higher than the values given above for standard scrubbers.

"Burn boxes" or oxidizers are used as treatment systems to incinerate contaminants during the manufacturing process. Advantages include complete destruction, relatively small size and low maintenance requirements. Energy cost is very high, however, since the system burns either hydrogen or (in some systems) natural gas, or uses electrical resistance heat, on a continuous basis. One additional problem is that, unless optics for the sensors are kept clean, higher and higher temperatures are generated, leading to nitrous oxide production. Older systems are especially prone to produce NO_x. Initial equipment costs are very high, approaching \$2,000/CFM. For these reasons, burn boxes are not recommended as a gas cylinder release treatment system.

Carbon absorption is an appropriate technology for treatment of some of the HPM gases typically found in wafer fab gas storage rooms. These systems absorb contaminants from the airstream and, eventually, become contaminated, requiring replacement. Units are typically of approximately 1000 CFM capacity and can be easily combined in parallel for higher flow rates. Initial equipment cost is moderate, at \$10/CFM, and energy cost is relatively low.

Maintenance is also minimal, generally limited to filter change out, and equipment space requirements are moderate. Drawbacks include a potential for fire due to carbon's combustibility and, ironically, its absorption ability. As "harmless" materials (dust, etc.) are absorbed, the carbon's ability to absorb any potential released gas is reduced. Consequently, frequent checks must be made to check system capacity. Replacement filters of the "bag-in/bag-out" type must be obtained from the manufacturer of the initial unit, since they are specially designed to fit. Finally, contaminated filters must be disposed of as hazardous waste.

4.7   AIR HANDLING SYSTEM ISOLATION

4.7.1   Several code sections apply to the need for fab air handling system isolation. They are as follows:

> **UFC 51.105(d)2   (Fabrication Areas) Ventilation Requirements - Separate Systems**
> "The return-air system of one fab area shall not connect to another system within the building."

> **UBC 911(b)3 Division 6 Occupancies, Fabrication Area - Ventilation**
> "...The exhaust air duct system of one fabrication area shall not connect to another duct system outside that fabrication area within the building..."

UFC 51.105(d)2 prohibits the return air system of a fab from being mixed with the return air from any other system outside the fab occupancy. It is acceptable to provide supply air from a fab make-up air handling system to another area if the supply air system is a 100 percent outside air unit. The supply ducts must be fire dampered if they pass through fire-rated walls in process. As a practical matter, it is not energy efficient to use the clean, humidified/dehumidified make-up air (which is necessary for the fab clean spaces) to service areas which do not require this type of conditioning. It has generally proven cost-effective and safe to provide separate air handling for non-fab areas and exit corridors.

The exit corridors may not be used as a return air space for another area. As a practical matter, when a fab area adjacent to the exit corridor is at positive pressure to the corridor (to control cleanliness) some air will be conveyed by the corridor. Serious thought needs to be given to the pressure hierarchy of the exit corridor/fab relationship during a smoke or other hazardous event (such as a toxic gas leak). Life safety concerns would dictate a higher pressure in the exit corridor than the fab under these conditions.   Production concerns require higher pressure in the fab.   Cost concerns may dictate not changing the hierarchy during a hazardous event.

UBC 911(b)3 prohibits the connection of an exhaust system in a fab to another duct system outside the fab area.   This has generally been interpreted to mean that the exhaust system must be fully contained within the fab area and must leave the fab area via the roof or an outside (non-fire-rated) wall. Note, however, that it is acceptable to connect the exhaust ducts for two separate fab areas together at a point outside the building. Such an interconnection of multiple fab exhaust systems might be desirable to allow cost effective installation of fume scrubbing or solvent abatement systems common to the fabs.

When an exit corridor is utilized to transport HPMs (as is allowed in existing fabs which are remodeled), the air handling system for the "service/exit" corridor must be completely isolated from the fab area air handling system.

Compliance with these provisions will limit the propagation of smoke or hazardous chemicals/gases from affected fabrication areas to exit corridors or adjacent spaces, thus, enhancing life safety.

4.8   SMOKE AND FIRE BARRIERS

4.8.1   Several code sections apply to this subject. They are as follows:

UBC 503(a)   Classification of All Buildings By Use   - Mixed Occupancy

"When a building is used for more than one occupancy purpose, each part of the building comprising a distinct 'Occupancy,' as defined in Chapters 5 through 12, shall be separated from any other occupancy as specified in Section 503(d)."

UBC 503(c)4   Classification of All Buildings By Use - Types of Occupancy Separations

"A 'one-hour' fire-resistive occupancy separation shall be of not less than one-hour fire-resistive construction.   All openings in such separations shall be protected by a fire assembly having a one-hour fire-protection rating."

UBC503(d)   Classification of All Buildings By Use - Fire Ratings for Occupancy Separations

"Occupancy separations shall be provided between the various groups and divisions of occupancies as set forth in Table No. 5-B."

UBC 911(b)1   (Division 6 Occupancies) Fabrication Area - Separation

"Fabrication areas, whose size is limited by the quantity of HPM permitted by the Fire Code, shall be separated from each other, from exit corridors and from other parts of the building by not less than one (1) hour fire-resistive occupancy separation.

Exceptions:

1   Doors within such occupancy separation, including doors to corridors, shall be only self-closing fire assembles having a fire protection rating of not less than 3/4 hour ..."

UFC 51.106(c)2C (Workstations within Fabrication Areas) Exhaust Ventilation - Duct Systems

"Exhaust duct systems penetrating occupancy separations shall be contained in a shaft of equivalent fire resistive construction. Ducts shall <u>not</u> penetrate area separation walls. Fire dampers shall <u>not</u> be installed in exhaust ducts."

UBC 1706(a) Requirements Based on Type of Construction - Shaft Enclosures

"Openings through floors shall be enclosed in a shaft enclosure of fire-resistive construction having the time period set forth in Table No. 17-A ... See occupancy chapters for special provisions."

(Note: See Chapter 2 of this book for discussion of multi-level fab.)

UBC 911 states the requirement for one-hour walls between fab areas and other fab areas and corridors. Penetrations of these walls with supply or return-air ducts must be protected with fire dampers. Process exhaust ducts must not penetrate the walls because they cannot be fire-dampered.

UBC 906 requires draft-stopping penetrations between floors within a fabrication area. This provision recognizes the configuration of a "multiple-level" fab which may be up to three stories high and have air handling on several levels. The "multiple-level" fab is considered one fire zone vertically; therefore, fire dampers are not required. Draft stops are provided to prevent the spread of smoke between levels.

Smoke dampers, or combination smoke/fire dampers, must be provided at all penetrations of ducts through a smoke partition. The requirements for smoke partitions are defined in the UBC and relate to the need for separation of various occupancies and "buildings." Duct penetrations of fire-rated corridor walls must be provided with combination fire/smoke dampers (UBC 4306(j)5). Our recommendation for the smoke damper actuator is to use a normally closed, pneumatic motor to achieve a fail-safe operating mode.

Exhaust ducts may not penetrate area separation walls (i.e., between two fab occupancies) and are subject to other restrictions. (*See "Air Handling System Isolation" above*).

4.8.2   Smoke Control Systems:

While smoke control air handling systems are not specifically required by the various codes for hazardous occupancies, your insurance carrier may have specific requirements for them.

One insurance carrier used by many industrial facilities requires a smoke exhaust (and corresponding make-up air system) that provides a ventilation rate of 4 CFM per square foot of area. This flow rate may be considerably higher than the normal ventilation rate. Therefore, a distinct system may be necessary to accommodate the requirement. Alternatively, the normal exhaust and make-up air system may need to be provided with smoke dampers to shift more than the normal flow rate to the smoke zone.

The method of implementing smoke control systems must be carefully designed to ensure it will be "fail safe" in the event of a power failure. If smoke dampers are used in conjunction with the normal ventilation system, they must be designed such that they cannot interfere with proper operation of the ventilation system under normal circumstances.

4.9   TEMPERATURE CONTROL

Control of temperature in areas where temperature-sensitive hazardous materials are stored or used is mandated by UFC 80.301(t)3. When temperature control is required to ensure the materials do not become unstable, the systems which control temperature must be provided with emergency power.

SUMMARY

MECHANICAL HEATING, VENTILATING AND AIR CONDITIONING SYSTEMS (4.0)

EXHAUST VENTILATION SYSTEMS (4.3)

Hazardous Area Ventilation Requirements (Including Workstations):

- Rooms, areas or spaces in which explosive, corrosive, combustible, flammable or highly toxic dusts, mists, fumes, vapors or gases are or may be emitted and shall be mechanically ventilated as required by the Fire Code and the Mechanical Code.

- Emissions generated at workstations shall be confined to the area in which they are generated.

- Exhaust ventilation shall be provided to produce not less than one (1) cubic foot per minute, per square foot floor area, and shall be in accordance with Section 79.805(e).

- A "fabrication area" is an area within a Group H-6 occupancy in which there are processes involving HPMs and may include ancillary rooms such as dressing rooms, offices, etc.

- We believe it is a good practice to provide general, or process, exhaust in all areas classified as H-6, at a rate of not less than 1 CFM/SF, or 6 air changes per hour, due to the possible migration of potentially hazardous fumes.

- Flammable/explosive vapors must be exhausted as directly as feasible.

- Separate duct/fan systems shall be provided for incompatible vapors and air streams:
    - Corrosive or general scrubbed exhaust.
    - Solvent/Hydrocarbon exhaust.
    - Toxic exhaust, treated.
    - General or "heat" exhaust.

- UMC 1105(a) requires that ducts which convey explosives or flammable vapors (toxics should be included in this group) shall extend directly to the exterior of the building, without entering other spaces. We believe it is <u>unacceptable</u> to connect gas cabinet exhaust to a workstation or any other exhaust ventilation system.

- Mixtures within work areas where contaminants are generated shall be diluted below 25 percent of their lower explosive limit or lower flammability limit.

- Dampers provided to balance air flow shall be provided with securely fixed minimum position blocking devices to prevent restricting flow below the required volume or velocity.

- Make-up air shall be provided to replenish air exhausted by the ventilation system. Make-up air intakes shall be located so as to avoid recirculation of contaminated air within enclosures.

- The location and means of discharging air to the atmosphere must be carefully considered. High velocity (greater than 3,500 FPM) vertical discharge is generally required to effect dilution of the potentially hazardous air stream with the ambient air.

- As a practical matter, the amount of make-up air required to replace the exhaust air and <u>pressurize</u> the space may be significantly higher than the exhaust flow.

■   Mechanical exhaust ventilation shall be provided in storage rooms at the rate of not less than one (1) cubic feet per minute per square foot of floor area or six (6) air changes per hour.

■   Exhaust ventilation shall be arranged to consider the density of the potential fumes or vapors released. For fumes or vapors that are heavier than air, exhaust shall be taken from a point within 12 inches of the floor.

■   The exhaust ventilation system should be connected to emergency power supply to ensure continuous operation.

■   Storage of toxic or highly toxic compressed gas cylinders shall be within ventilated gas cabinets, exhausted enclosures or within a ventilated separate gas storage room. Gas cabinets shall be operated at negative pressure in relation to the surrounding area, and shall be provided with self-closing limited access ports. The average velocity of ventilation shall be not less than 200 feet per minute.

EMERGENCY VENTILATION AND OPERATION (4.5)

♦   Emergency power is required to operate the exhaust ventilation and treatment systems in H-6 occupancies and H-2, H-3 or H-7 facilities where toxic or highly toxic gas is used. Emergency power must be on-line within ten seconds of failure of the primary power source. The start-up of legally required standby power can be delayed for up to 60 seconds after a power failure.

♦   We believe that fab area pressurization should be maintained to control cleanliness. Therefore, it is recommended that each facility consider providing outside make-up air in excess of exhaust air quantities on emergency power.

♦ It is prudent, if not mandatory, to provide emergency power, or preferably an uninterruptible power supply (UPS) for all controls which operate the life safety related equipment.

■ <u>Except for exhaust systems</u>, at least one manually-operated remote control switch that will shut down the fab area ventilation system shall be installed at an approved location outside the fabrication area.

■ The manual shutdown switch can be a source of potential hazard. Consider placing the switch in the "emergency control station" where only qualified personnel have access.

EXHAUST TREATMENT AND SEPARATION OF REACTIVE MATERIALS (4.6)

♦ Separation: The Codes require separation of incompatible air streams such as acid and solvent exhaust.

♦ Good engineering practice and pollution abatement regulations generally require separation of solvent and corrosive exhaust.

♦ Treatment: The Uniform Fire Code (UFC), Section 80.303(a)6D requires a "treatment system" to process the exhaust discharged from gas cabinets, exhausted enclosures or separate storage rooms.

■ Dilution is rarely effective as the sole means of treatment of an unrestricted cylinder release, due to the great amount of exhaust flow required.

■ If a restricted flow orifice (RFO) is placed in the cylinder, the rate of release during a catastrophic event is reduced substantially, therefore, dilution becomes a more viable treatment method.

■ The traditional method of removing acids and fumes from air streams has involved wet scrubbing, using water or a water-based chemical solution.

■   Treatment of toxic metal hydrides present special problems in that the contaminants are generally not removed using standard chemistries for a scrubbing process. Generally, oxidation-reduction systems are employed with scrubbers using potassium permanganate ($KMnO_4$).

■   Burn boxes are used as treatment systems to incinerate contaminants or pyrophorics which are not reacted during the manufacturing process.

■   Carbon absorption may be an appropriate technology for treatment of some of the HPM gases typically found in wafer fab gas storage rooms (particularly "metal hydrides").

AIR HANDLING SYSTEM ISOLATION (4.7)

◆   The return-air system of one fab area shall not connect to another system within the building.

◆   The exhaust air duct system of one fabrication area shall not connect to another duct system outside that fabrication area within the building.

◆   It is acceptable to provide <u>supply</u> air from a fab "primary" air handling system to another area <u>if the supply air system is a 100 percent outside air</u> unit.

◆   The exit corridors may not be used as a return air space for another area.

◆   It is acceptable to connect the exhaust ducts for two separate fab areas together at a point <u>outside</u> the building.

◆   Compliance with these provisions will limit the propagation of smoke or hazardous chemicals/gases from affected fabrication areas to exit corridors or adjacent spaces and enhance life safety.

SMOKE AND FIRE BARRIERS (4.8)

♦   When a building is used for more than one occupancy purpose, each part of the building comprising a distinct 'Occupancy,' shall be separated from any other occupancy.

♦   Fabrication areas, whose size is limited by the quantity of HPM permitted by the Fire Code, shall be separated from each other, from exit corridors and from other parts of the building by not less than one (1) hour fire-resistive occupancy separation.

♦   Exhaust duct systems penetrating occupancy separations shall be contained in a shaft of equivalent fire resistive construction. Ducts shall <u>not</u> penetrate area separation walls. Fire dampers shall <u>not</u> be installed in exhaust ducts.

♦   UBC 906 requires draft-stopping penetrations between floors within a fabrication area. This provision recognizes the configuration of a "multiple-level" fab which may be up to three stories high and have air handling on several levels.

♦   Smoke dampers, or combination smoke/fire dampers, must be provided at all penetrations of ducts through a smoke partition.

♦   Duct penetrations of fire-rated corridor walls must be provided with combination fire/smoke dampers (UBC 4306(j)5).

5

Fire Suppression

5.1 GENERAL:

Fire sprinklers are the major component of the building fire suppression system. All Group H, Division 1, 2, 3, 6 and 7 occupancies are required to be sprinklered.

UFC 10.507(f)1 Required Installation of Automatic Fire-Extinguishing Systems - Group H Occupancies - Divisions 1, 2, 3, and 7
"An automatic fire extinguishing system shall be installed in Group H, Division 1, 2, 3 and 7 Occupancies."

UFC 10.507(f)3 Required Installation of Automatic Fire-Extinguishing Systems - Group H Occupancies - Division 6
"An automatic fire extinguishing system shall be <u>installed throughout buildings</u> containing Group H, Division 6 occupancies. The design of the sprinkler system shall not be less than ... as follows:

Location	Occupancies Hazard Classification
Fabrication Areas	Ordinary Hazard Group 3
Service Corridors	Ordinary Hazard Group 3
Storage rooms without dispensing	Ordinary Hazard Group 3
Storage rooms with dispensing	Extra Hazard Group 2
Exit Corridors*	Ordinary Hazard Group 3

*   When the design area of the sprinkler system consists of a corridor protected by one row of sprinklers, the maximum number of sprinklers that need be calculated is 13."

These same requirements are stated in UBC Section 3802(f) and (more specifically) require sprinkler systems to be not less than that required under UBC Standard No. 38-1.

UFC 80.301(p) (Hazardous Materials) Storage - Fire extinguishing systems

"Unless exempted, ... indoor storage areas and storage buildings shall be protected by a sprinkler system.   The design ... shall be not less than ... for Ordinary Hazard Group 3 with a minimum design area of 3000 square feet. See UBC Standard 38-1.   Where the materials or storage arrangement require a higher level of ... protection, ... the higher level ... shall be provided. . ."

UFC 80.401(r) Dispensing, Use and Handling - Fire extinguishing systems

"Indoor rooms or areas in which hazardous materials are dispensed or used shall be protected by a sprinkler system. Sprinkler system design ... shall be not less than ... for Ordinary Hazard Group 3 with a minimum design area of 3000 square feet.   See UBC Standard 38-1.   Where the materials or storage arrangement require a higher level of ... protection, ... the higher level ... shall be provided. . ."

5.2   REVISIONS AND UPGRADES

UFC 10.504(a) Installation and Maintenance of Fire-Protection and Life-Safety Systems

"Sprinkler systems, fire hydrant systems . . .   shall be maintained in an operative condition at all times and shall be replaced or repaired where defective . . . Such systems shall be extended, altered or augmented as necessary to maintain and continue protection whenever any building so equipped is altered, remodeled or added to . . ."

Note the provision that applies to remodeling (or upgrading the facility from B-2 to "H" occupancy).   Whenever alterations, remodeling or additions are done to any part of the building, the fire sprinkler system must be augmented to provide proper coverage. Hence, it is important to design flexibility and spare capacity for extra tees and drops in the system to keep pace with the ever-changing needs of an industrial facility. Retrofit/rework of manufacturing areas is a relatively fast-paced procedure. Sprinkler systems must be altered as work is done in order to assure coverage to all portions of the fab.

5.3   H-6 OCCUPANCIES

5.3.1   The Uniform Fire Code requires complete sprinklering of the building in which a Group H, Division 6 Occupancy exists. The design of the sprinkler protection system must be in accordance with UBC Standard No. 38-1 with the hazard classifications as outlined above. (*Refer to* **Figure 5.1** *for illustration of sprinkler head integrated into a cleanroom ceiling grid.*)

Figure 5.1   FIRE SPRINKLER HEAD Integrated in Cleanroom Ceiling Grid. (Photo courtesy of Daw Technologies, Inc.)

5.3.2   Most (but not all) Authorities Having Jurisdiction, when adopting the UBC, modify the requirement for sprinkler system compliance with UBC Standard 38-1 such that compliance with applicable NFPA codes is the standard. The codes most frequently cited are:

1.   NFPA 13 - Installation of Sprinkler Systems,
2.   NFPA 14 - Standpipe and Hose Systems,
3.   NFPA 24 - Private Fire Service Mains and their Appurtenances and
4.   NFPA 318 - Protection of Cleanrooms.

While NFPA 13, 14, and 24 have been in existence for many years, NFPA 318 is new and covers many specialized installations found within cleanroom facilities.

NFPA 318 2-1.2.1   (Protection of Cleanrooms) Fire Protection - Automatic Sprinkler Systems
"Automatic sprinklers for cleanrooms or clean zones shall be in accordance with NFPA 13 ... and shall be hydraulically designed for a density of 0.20 gpm/square feet over a design area of 3000 square feet. "

This density <u>exceeds</u> the Ordinary Hazard Group 3 occupancy design density (per NFPA 13) of 0.18 gpm/square feet over 3000 square feet.

NFPA 318 2-1.2.2   (Protection of Cleanrooms) Fire Protection - Automatic Sprinkler Systems
"Approved quick response sprinklers shall be utilized for sprinkler installations within downflow airstreams in cleanrooms and clean zones. "

In a downflow cleanroom, the sprinkler head is continuously being cooled by air at 65°F. to 70°F. at a velocity of 60 to 100 feet per minute. Therefore, a standard head will be slower to respond to a fire situation than may be desirable.   The use of quick-response sprinklers, while still delayed in opening by the downward airflow, will respond to a small fire quicker than conventional sprinklers.

5.3.3   In addition to code compliance, many hazardous occupancies are insured through carriers (such as Factory Mutual (FM) or Industrial Risk Insurers (IRI)) who have additional design and construction requirements, in order to reduce the risk of fire and other liabilities. Necessary first steps in rational design of sprinkler systems are determination of:

1.   What standards apply for sprinklers, and

2.   What, if any, additional insurance carrier requirements are in effect?

5.3.4   Plenums and Interstitial Spaces:

NFPA 318 2-1.2.4   (Protection of Cleanrooms) Fire Protection - Automatic Sprinkler Systems
"Automatic sprinkler protection shall be designed and installed in the plenum and interstitial space above cleanrooms in accordance with NFPA 13 ... for a density of 0.20 gpm/square feet over a design area of 3000 square feet."

5.3.5   Workstations:

All workstations within the fab must be installed with fire sprinkler protection. A "workstation" is a defined space or independent piece of equipment using hazardous production materials within a fabrication area (i.e., production equipment, hoods and related enclosures.)

UFC 51.106(d)1-3 (Semiconductor Fabrication Facilities Using HPM) Workstations within Fabrication Areas - Fire Protection
"Sprinkler coverage of the horizontal surface at any workstation shall not be obstructed. A sprinkler shall be installed within the exhaust duct connection of workstations of combustible construction . . .

Exceptions:

1   Approved alternate fire-extinguishing systems are allowed. Activation of such systems shall deactivate the related processing equipment.

2   Process equipment which operates at temperatures exceeding 500° C and which is provided with automatic shutdown capabilities for HPM.

3   Exhaust ducts 10 inches or less in diameter from flammable gas storage cabinets that are part of a workstation."

Workstations constructed of combustible materials containing HPMs must contain a sprinkler head within four feet of the exhaust duct connection. This helps to prevent plastic, which, in a fire, becomes a flammable liquid, from being drawn into the duct, potentially transporting the fire.

NFPA 318, in addition to requiring automatic sprinkler protection of the horizontal surface (paragraph 2-1.7.1), requires sprinkler protection in the exhaust transition piece of combustible work stations (paragraph 2-1.7.2). Appendix A of NFPA 318 illustrates (in Fig. A-2-1.2.7) various arrangements of a wet bench work station, the associated fume exhaust ductwork, and possible location(s) of fire protection devices. (*Refer to* **Figure 5.2.**)

UFC 51.106 and NFPA 318 also have significant impact on the retrofit of buildings with combustible workstations due to the installation and inspection requirements of sprinklers in the ductwork.

5.4   STORAGE OCCUPANCIES

5.4.1   Hazardous material storage rooms classified as H-1, H-2, H-3 and H-7 require complete sprinkler coverage in accordance with UFC 10.507(f), UFC 79.806 and UBC Section 3802(f).

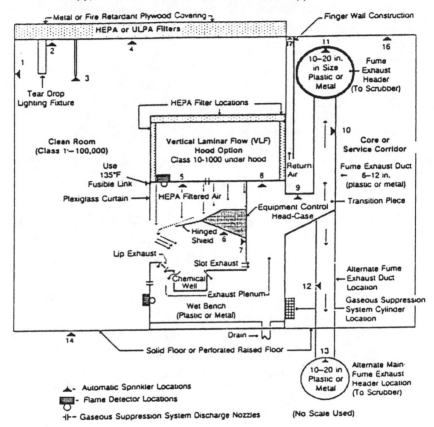

Figure 5.2   EXAMPLE OF SPRINKLER LOCATIONS FOR COMBUSTIBLE WORK STATIONS. (NFPA 318, Figure A-2-1.2.7   Reprinted with permission from NFPA 318: *Protection of Cleanrooms,* Copyright c 1992, National Fire Protection Association, Quincy, MA 02269.   This reprinted material is not the complete and official position of the National Fire Protection Association,   on the referenced subject which is represented only by the standard in its entirety.)

5.4.2   In accordance with UFC 79.805 (pertaining to areas for the Use, Dispensing and Mixing of Flammable and Combustible Liquids), "Use, dispensing and mixing room shall be classified as Group H, Division 2 or 3 Occupancies as required by the Building Code..."

> **UFC 79.806 Flammable and Combustible Liquids - Fire Protection**
> "Fire protection shall be provided in accordance with Article 10."

5.4.3   In accordance with UFC Table 10.510-A, Class II standpipes are required in H-1, H-2 and H-3 occupancies of non-sprinklered buildings greater than 20,000 square feet per floor. This type of storage occupancy is rarely larger than 20,000 SF in a wafer fabrication facility.

5.4.4   Rooms used for storage of HPMs (which must be sprinklered) must also have provisions for drainage, containment and secondary containment of the spills <u>and</u> fire protection water equal to a full discharge of the sprinklers for a period of 20 minutes (UFC 80.301(l) 3 and 4 and NFPA 318 5-1.1.2.). When dealing with small rooms, it is essential to consider actual sprinkler head flow rate in addition to design density, because the actual flow rate may be significantly higher. The actual flow rate can be calculated knowing the available pressure at the head and the orifice flow coefficient of the heads.

> **UFC 51.110(b)7A (Storage of HPM within Buildings) Special Provisions - Spill Control and Drainage Control**
> "Hazardous production material storage rooms for HPM liquids shall have spill control and drainage control in accordance with Section 80.301(l). Hazardous production material flammable liquid drains shall be separated from other hazardous production material liquid drains..."

(More on this subject is provided in Chapters 2 and 3 of this book.)

5.4.5   The requirements of UFC Article 80 concerning fire suppression for facilities used to store HPMs in excess of the exempt amounts include the following references.   Be sure to use these in context with the material referenced in the applicable section.

> **UFC 80.301(p) Hazardous Materials - Fire-extinguishing Systems**
>
> "Unless exempted or otherwise provided for in Sections 80.302 through 80.315, indoor storage areas and storage buildings shall be protected by an automatic sprinkler system. The design of the sprinkler system shall be not less than that required by the Building Code for Ordinary Hazard Group 3 with a minimum design area of 3,000 square feet. See UBC Standard No. 38-1.   Where the materials or storage arrangement require a higher level of sprinkler system protection in accordance with nationally recognized standards, the higher level of sprinkler system protection shall be provided.
>
> > **Exception:**   Approved alternate automatic fire-extinguishing systems are allowed. "
>
> **UFC 80.303(a)3A-B (Toxic and Highly Toxic Compressed Gases) Indoor Storage - Fire-Extinguishing System**
>
> "In addition to the requirements of Section 80.301(p), the following requirements shall apply:
>
> A   Gas cabinets or exhausted enclosures for the storage of cylinders shall be internally sprinklered.
>
> B   Alternate fire-extinguishing systems shall not be used for either storage areas, gas cabinets or exhausted enclosures. "
>
> **UFC 80.307(b) Organic Peroxides - Exterior Storage**
>
> (See text of the UFC if such storage is applicable).

UFC 80.310(a)8 (Water Reactive Solids and Liquids) Indoor Storage - Fire-Extinguishing Systems
"When Class 3 (water reactive) materials are stored in areas protected by an automatic fire-sprinkler system, the materials shall be stored in closed watertight containers."

UFC 80.312(b)3A-B (Highly Toxic and Toxic Solids and Liquids) Exterior Storage - Fire-Extinguishing Systems
"Exterior storage of highly toxic solids and liquids shall be in fire-resistive containers or shall comply with one of the following:

A   The storage area shall be protected by an automatic, open head, deluge fire-sprinkler system of the type and density specified in the Building Code (see UBC Standard No. 38-1), or

B   Storage shall be located under a canopy of noncombustible construction, with the canopied area protected by an automatic fire-sprinkler system of the type and density specified in the Building Code. See UBC Standard No. 38-1. Such storage shall not be considered indoor storage."

UFC 80.401(r) Dispensing, Use and Handling - Fire-Extinguishing Systems
"Indoor rooms or areas in which hazardous materials are dispensed or used shall be protected by an automatic fire-extinguishing system. Sprinkler system design shall be not less than that required by the Building Code for Ordinary Hazard, Group 3, with a minimum design area of 3,000 square feet. See UBC Standard No. 38-1. Where the materials or storage arrangement require a higher level of sprinkler system protection in accordance with nationally recognized standards, the higher level of sprinkler system protection shall be provided.

       Exception: Approved alternate automatic fire-extinguishing systems are allowed."

UFC 80.402(b)2D Indoor Dispensing and Use - Fire-Extinguishing System

"In addition to Section 80.401(r), laboratory fume hoods and spray booths where flammable materials are dispensed or used shall be protected by an automatic fire-extinguishing system."

UFC 80.402(c)3 Exterior Dispensing and Use - Fire-Extinguishing System

"Flammable hazardous materials dispensing or use areas located within 50 feet of either a storage area or building, and vehicle loading racks where flammable hazardous materials are dispensed, shall be protected by an approved fire-extinguishing system."

UFC 80.402(c)6A-C Exterior Dispensing and Use - Fire Access Roadways and Water Supply

A   Fire access roadways and approved water supplies shall be provided for exterior dispensing or use area in accordance with this subsection.

B   Fire apparatus access roadways shall be provided to within 150 feet of all portions of an exterior dispensing or use area. Such access roadways shall comply with Article 10, Division II.

C   An approved water supply shall be provided. Fire hydrants or other approved means capable of supplying the required fire flow shall be provided to within 150 feet of all portions of an exterior dispensing or use area. The water supply and fire hydrants shall comply with Article 10, Division IV.

UFC 80.402(c)8A (Exterior Dispensing and Use) Special Requirements for Highly Toxic or Toxic Compressed Gases - Ventilation and Storage Arrangement

"When cylinders or portable containers are used out-of-doors, gas cabinets or a locally exhausted enclosure shall be provided."

UFC 80.402(c)8E   (Exterior Dispensing and Use) Special Requirements for Highly Toxic or Toxic Compressed Gases - Fire-Extinguishing System
"Gas cabinets and exhausted enclosures shall be internally sprinklered."

5.5   EXHAUST DUCTS CONTAINING FLAMMABLE VAPORS

5.5.1   Exhaust ducts which contain flammable vapors must contain fire protection per the Uniform Mechanical Code section 1107(g) and the first portion of Article 51.106(c)2D of the Uniform Fire Code.

UMC 1107(g) (Ventilation Systems and Product-conveying Systems) Product-conveying Ducts - Fire Protection
"Sprinklers or other fire-protection devices shall be installed within ducts having a cross-sectional dimension exceeding 10 inches when the duct conveys flammable vapors or fumes. Sprinklers shall be installed at 12-foot intervals in horizontal ducts and at changes in direction. In vertical runs, sprinklers shall be installed at the top and at alternate floor levels."

Section 1102 of the UMC defines flammable vapor or fumes as "... flammable constituents in air that exceed 10 percent of its lower flammability limit (LFL)." For example, a gas that is flammable in air from 15 percent to 80 percent concentration would have to be diluted to 1.5 percent concentration (10 percent of 15 percent) before it would be considered non-flammable.

UFC 51.106(c)2D(i-iii)   (Exhaust Ventilation) Duct Systems - Fire Protection
"Exhaust ducts shall be internally sprinklered when all of the following conditions apply:
(i)   When the largest cross-sectional diameter is equal to or greater than 10 inches,
(ii)   The ducts are within the building, and

(iii)   The ducts are <u>conveying gases or vapors in a flammable range</u>. . ."

5.5.2 In order to meet the above criteria, sprinklers must be installed in flammable exhaust ducts equal to or greater than ten (10) inches cross-sectional diameter that are within the building. These include all combined acid and solvent exhaust duct systems. The "alternative" to sprinklering (i.e., dilution to less than 10 percent LFL) is not recommended, since a failure of the exhaust system fan, damper closure, etc. could compromise the dilution. It is recognized that the typical dilution rate in a large exhaust system <u>probably</u> could be shown to result in a mixture which is below the lower flammable limit.   However, it is not considered good risk management to eliminate duct sprinklers for several reasons:

1.   The cost of sprinklers is <u>relatively</u> low.
2.   The potential damage of a duct fire is extremely high.
3.   Anomalies in operation could lead to higher than anticipated concentrations; therefore, concentrations within the flammable range.

NFPA 318 2-1.2 covers design criteria for sprinkler systems for ductwork. (*These criteria are covered in Section 5.6 below.*)

In processes where evacuated chambers are created, oil lubricated vacuum pumps are commonly used. If the pump is prone to misting, oil vapors could be entrained and condense on the exhaust duct walls. Duct sprinklers in these applications must be evaluated prudently since, if combustion occurs, spraying the oil-lined duct with water could serve to spread the fire. The need for observation and periodic cleaning of such duct systems cannot be overstated.

5.6   EXHAUST DUCTS CONTAINING CORROSIVE VAPORS

5.6.1 Unless exempted as noted below, exhaust ducts which contain corrosive vapors are considered to be "product conveying" (i.e., carrying fumes, smoke, mists, fogs, vapors, noxious or toxic gases or air above 250°F., per UMC Section 1102) and require fire protection per Uniform Mechanical Code Chapter 11 and the last portion of Article 51.106(c)2D of the Uniform Fire Code.

UMC 1107(a) (Ventilation Systems and Product-conveying Systems) Product-conveying Ducts - Classification

"Product-conveying ducts shall be classified according to their use as follows...

...Class 5.   Ducts conveying corrosives,   such as acid vapors."

UMC 1107(b) (Ventilation Systems and Product-conveying Systems) Product-conveying Ducts - Materials

"Material... shall be of metal.

> **Exceptions...** Ducts serving a Class 5 system may be constructed of approved nonmetallic material when the corrosive characteristics of the material being conveyed make a metal system unsuitable and when the mixture being conveyed is nonflammable...
>
> Approved nonmetallic material shall be either a listed product having a flamespread index of 25 or less and a smoke-developed rating of 50 or less on both inside and outside surfaces without evidence of continued progressive combustion, or shall have a flame-spread index of 25 or less, and shall be installed with an automatic fire sprinkler protection system inside the duct."

UFC 51.106(c)2D   (Exhaust Ventilation) Duct Systems - Fire Protection

"Combustible nonmetallic ducts whose largest cross-sectional diameter is equal to or greater than 10 inches shall be internally sprinklered.

> **Exceptions:**
>
> 1   Ducts listed for nonsprinklered applications.
>
> 2   Ducts not more than 12 feet in length installed below ceiling level."

Acceptability of a duct for use without sprinklers must be confirmed with the Authorities Having Jurisdiction over the Project.

5.6.2   NFPA 318 2-1.2 covers design criteria for ductwork sprinkler systems, including density/coverage, zoning, drainage, and inspection/maintenance requirements.

> **NFPA 318: 2-1.2.5.1 (Protection of Cleanrooms) Fire Protection - Automatic Sprinkler Systems**
> "Sprinklers installed in duct systems shall be hydraulically designed to provide 0.5 gpm over an area derived by multiplying the distance between the sprinklers in a horizontal duct by the width of the duct.   Minimum discharge shall be 20 gpm per sprinkler from the 5 hydraulically most remote sprinklers.   Sprinklers shall be spaced a maximum of 20 feet apart horizontally and 12 feet apart vertically."
>
> (12   20
>
> **NFPA 318:   2-1.2.5.2   Fire Protection - Automatic Sprinkler Systems**
> "A separate indicating control valve shall be provided for sprinklers installed in ductwork."
>
> **NFPA 318:   2-1.2.5.3   Fire Protection - Automatic Sprinkler Systems**
> "Drainage shall be provided to remove all sprinkler water discharged in ductwork."

> **NFPA 318: 2-1.2.5.4 Fire Protection - Automatic Sprinkler Systems**
> "Where corrosive atmospheres exist, duct sprinklers and pipe fittings shall be manufactured of corrosion resistant materials or coated with approved materials."

> **NFPA 318: 2-1.2.5.5 Fire Protection - Automatic Sprinkler Systems**
> "The sprinklers shall be accessible for periodic inspection and maintenance."

These criteria apply to sprinkler systems for both corrosive and flammables-conveying ducts. Note the requirements for zoning of the sprinkler system, drainage of the duct and access openings in the duct to facilitate inspection.

5.7   GAS CABINETS

All gas cabinets located within the H-6 fab areas must contain sprinklers. In H-2 or H-7 storage rooms, only cabinets containing pyrophoric materials must be sprinklered (see exception to UFC 51.107(b)). If the design of the gas cabinets allows venting of the HPM or process gas into the cabinet exhaust duct, the duct is considered to be product-conveying and should be provided with sprinklers even if not required by code. In addition, to ensure flexibility, the author recommends sprinkling of all cabinets, regardless of the gas which a cabinet will store at the outset of design.

> **UFC 51.107(b)1 Storage and Dispensing of HPM within Fabrication Areas - Requirements for HPM Gases**
> "Cabinets used for the containment of HPM gases shall be in accordance with Section 51.107(a) and this subsection. Gas cabinets containing HPM gases shall be internally sprinklered.

> **Exception:** Sprinklers shall not be required in gas cabinets that are located within an HPM storage room other than those cabinets used to contain pyrophoric gases. . ."

5.8   EXIT CORRIDORS

5.8.1   Exit corridors require general sprinkler protection of Ordinary Hazard Group 3, identical to the fab.

> **UFC 51.108(b)9 (Handling of HPM within Exit Corridors) Existing Buildings - Fire Protection**
> "Sprinkler protection shall be designed in accordance with the Building Code as required for Ordinary Hazard Group 3. See UBC Standard No 38-1.   When the design area of the sprinkler system consists of one row of sprinklers in the corridor, the maximum number of sprinklers to be calculated need not exceed 13."

5.8.2   Like the general fab, the Authorities Having Jurisdiction for many areas require use of NFPA standards in lieu of UBC Standard No. 38-1.   Since cleanroom exit corridors are often cleanrooms in and of themselves, (e.g., Class 10,000 corridors), where adopted, NFPA 318 - Protection of Cleanroom requirements apply to exit corridors.

5.8.3   Piping Chases:

If HPM piping is installed in or over the exit corridors, fire sprinklers must be installed in the piping chase, unless the chase is less than six inches in dimension.

UBC 911(f)2A-F (Division 6 Occupancies) Piping and Tubing - Installations in Exit Corridors and above Other Occupancies.

". . . . Hazardous production material piping and tubing may be installed within the space defined by the walls of exit corridors and the floor or roof above or in concealed spaces above other occupancies under the following conditions:

A   Automatic sprinklers shall be installed within the space unless the space is less than 6 inches in least dimension."

B-F   (See code for additional conditions)

This requirement applies in all cases, even if the construction of the chase utilizes non-combustible material.

5.9   SERVICE CORRIDORS

Sprinkler design conforming to Ordinary Hazard Group 3 is required in service corridors (used for transporting HPM materials in an H-6 occupancy).

UFC 51.109(a)1-6   Handling of HPM within Service Corridors - Transportation Criteria

"A service corridor shall be provided when necessary to transport HPMs to and from an HPM storage room, or from the outside of a building to the perimeter wall of the fabrication area.   Service corridors shall not be used as a required exit corridor.   Service corridors used for transporting HPM shall be in accordance with this subsection.

2-5   (See code for additional conditions)

6   Fire protection sprinklers shall be installed in accordance with Section 51.108(b)8."

The sprinkler design criteria for the fab, exit corridors and service corridors are all Ordinary Hazard Group 3.   Because of this, the overall sprinkler design is not affected by corridor arrangements.

5.10   ALTERNATE SUPPRESSION SYSTEMS

5.10.1   In certain applications, water may not be the most effective or prudent means to extinguish fires.   In fact, in special cases, it could actually transport fire.   Hence, the Uniform Fire Code allows alternate suppression systems.

UFC 51.106(d)1-3 Workstations within Fabrication Areas - Fire Protection

"Exceptions:

1   Approved alternate fire-extinguishing systems may be used.   Activation of such systems shall deactivate the related processing equipment.

2   Process equipment which operates at temperatures exceeding 500° C. and which is provided with automatic shutdown capabilities for HPM.

3   Exhaust ducts. . . from flammable gas storage cabinets that are part of a workstation."

Hazardous Material Storage - Fire Protection

UFC 80.301(p) (Hazardous Materials) Storage - Fire Extinguishing Systems

"Exception: Approved alternate automatic fire extinguishing systems are allowed."

5.10.2   Halon and carbon dioxide are possible alternate suppression systems. Historically, halon has been the preferred alternate suppression medium (over carbon dioxide) due to its superior fire suppression capability with minimal adverse effects on occupants and firefighters. Due to recent changes in environmental regulations which ban the use or atmospheric discharge of certain CFCs (which may or may not include HCFCs such as halon), halon systems are now substantially less "popular." If used, system testing provisions should be altered to prevent any unnecessary halon release. Note, however, the word "approved" used in the exceptions in the code. The local Authorities Having Jurisdiction must review and approve all proposed substitutions. It may well be that, even with these provisions, halon systems are disallowed unless the owner can "prove" it is the only acceptable fire-fighting methodology. The owner may be required to provide the technical assistance necessary for this proof, under the following UFC provision:

UFC 2.302 Special Procedures - Technical Assistance

"To determine the acceptability of technologies, processes, products, facilities, materials, and uses attending the design, operation or use of a building or premises subject to the inspection of the department, the chief is authorized to require the owner or the person in possession or control of the building or premises to provide, without charge to the jurisdiction, a technical opinion and report. The opinion and report shall be prepared by a qualified engineer, specialist, laboratory or fire-safety specialty organization acceptable to the chief and the owner and shall analyze the fire-safety properties of the design, operation or use of the building or premises and the facilities and appurtenances situated thereon, to recommend necessary changes."

5.11   ZONING

Zoning of fire sprinkler systems should be arranged so that, when activated, the emergency signal sent to the life safety alarm and monitoring system can pinpoint the area of the fire. We recommend separate zoning for:

-       General Fab Area (one zone per fab per floor),
-       Interstitial Space or Plenum,
-       Mechanical Support Space,
-       Below Raised Floor,

- Solvent Exhaust Ducts,
- Acid Exhaust Ducts, and
- HPM Storage Rooms.

Flow and tamper switches must be installed on each zone.

SUMMARY

FIRE SUPPRESSION (5.0)

◆ Fire sprinklers are the major component of the building fire suppression system. All Group H, Division 1, 2, 3, 6 and 7 occupancies and their related storage areas are required to be sprinklered.

H-6 OCCUPANCIES (5.3)

◆ Most Authorities Having Jurisdiction, when adopting the UBC, modify the requirement for sprinkler system compliance with UBC Standard 38-1 to instead require compliance with applicable NFPA codes. The codes most frequently cited are:

- ■ NFPA 13 - Installation of Sprinkler Systems,
- ■ NFPA 14 - Standpipe and Hose Systems,
- ■ NFPA 24 - Private Fire Service Mains and their Appurtenances and
- ■ NFPA 318 - Protection of Cleanrooms.

◆ NFPA 318 is new and covers many specialized installations found within cleanroom facilities.

- ■ Automatic sprinklers for cleanrooms or clean zones shall be in accordance with NFPA 13 and shall be hydraulically designed for a density of 0.20 gpm/square feet over a design area of 3000 square feet.

- ■ Approved quick response sprinklers shall be utilized for sprinkler installations within downflow airstreams in cleanrooms and clean zones. The quick-response sprinklers, will respond to a smaller size fire quicker than conventional sprinklers.

- Automatic sprinkler protection shall be designed and installed in the plenum and interstitial space above cleanrooms in accordance with NFPA 13.

♦ Many hazardous occupancies are insured through carriers (such as Factory Mutual (FM) or Industrial Risk Insurers (IRI)) who have additional design and construction requirements, in order to reduce the risk of fire and other liabilities.

♦ Workstations: All workstations within the fab must be installed with fire sprinkler protection.

- Sprinkler coverage of the horizontal surface at any workstation shall not be obstructed. A sprinkler shall be installed within the exhaust duct connection of workstations of combustible construction.

- Workstations constructed of combustible materials containing HPMs <u>must</u> contain a sprinkler head within four feet of the exhaust duct connection.

- NFPA 318 requires sprinkler protection <u>in the exhaust transition piece</u> of combustible work stations.

STORAGE OCCUPANCIES (5.4)

♦ Hazardous material storage rooms classified as H-1, H-2, H-3 and H-7 require complete sprinkler coverage in accordance with UFC 10.507(f), UFC 79.806 and UBC Section 3802(f).

♦ Rooms used for storage of HPMs (which must be sprinklered) must also have provisions for drainage, containment and secondary containment of the spills <u>and</u> fire protection water equal to a full discharge of the sprinklers for a period of 20 minutes (UFC 80.301(l) 3 and 4 and NFPA 318 5-1.1.2.).

- Indoor storage areas and storage buildings shall be protected by an automatic sprinkler system   for Ordinary Hazard Group 3 with a minimum design area of 3,000 square feet.

- Gas cabinets or exhausted enclosures for the storage of toxic or highly toxic gas cylinders shall be internally sprinklered. Alternate fire-extinguishing systems shall not be used for either storage areas, gas cabinets or exhausted enclosures.

- When Class 3 (water reactive) materials are stored in areas protected by an automatic fire-sprinkler system, the materials shall be stored in closed watertight containers.

- Indoor rooms or areas in which hazardous materials are dispensed or used shall be protected by an automatic fire-extinguishing system for Ordinary Hazard, Group 3, with a minimum design area of 3,000 square feet.

EXHAUST DUCTS CONTAINING FLAMMABLE VAPORS (5.5)

◆ Exhaust ducts which contain flammable vapors must contain fire protection per the Uniform Mechanical Code Section 1107(g) and UFC Article 51.106(c)2D.

  - Sprinklers or other fire-protection devices shall be installed within ducts having a cross-sectional dimension exceeding 10 inches <u>when the duct conveys flammable vapors or fumes.</u>

  - Section 1102 of the UMC defines flammable vapor or fumes as "... flammable constituents in air that exceed 10 percent of its lower flammability limit (LFL)."

  - Exhaust ducts shall be internally sprinklered when all of the following conditions apply:
    - When the largest cross-sectional diameter is equal to or greater than 10 inches,
    - The ducts are within the building, and

The ducts are <u>conveying gases or vapors in a flammable range</u>.

♦ NFPA 318 2-1.2 covers design criteria for sprinkler systems for ductwork.

EXHAUST DUCTS CONTAINING CORROSIVE VAPORS (5.6)

♦ Exhaust ducts which contain corrosive vapors are considered to be "product conveying" and require fire protection per Uniform Mechanical Code Chapter 11 and UFC Article 51.106(c)2D.

- Combustible nonmetallic ducts whose largest cross-sectional diameter is equal to or greater than 10 inches shall be internally sprinklered.

♦ NFPA 318 2-1.2 covers design criteria for ductwork sprinkler systems, including density/coverage, zoning, drainage, and inspection/maintenance requirements.

- Sprinklers installed in duct systems shall be hydraulically designed to provide 0.5 gpm over an area derived by multiplying the distance between the sprinklers in a horizontal duct by the width of the duct.
- A separate indicating control valve shall be provided for sprinklers installed in ductwork.
- Drainage shall be provided to remove all sprinkler water discharged in ductwork.
- The sprinklers shall be accessible for periodic inspection and maintenance.

GAS CABINETS (5.7)

♦ All gas cabinets located within the H-6 fab areas must contain sprinklers. In H-2 or H-7 storage rooms, only cabinets containing pyrophoric materials must be sprinklered (see exception to UFC 51.107(b)).

♦   Sprinkling of <u>all</u> cabinets, regardless of the gas which a cabinet will store at the outset of design is <u>recommended</u>.

EXIT AND SERVICE CORRIDORS (5.8 and 5.9)

♦   Exit and service corridors require general sprinkler protection of Ordinary Hazard Group 3, similar to the fab.

♦   Piping Chases: If HPM piping is installed in or over the exit corridors, fire sprinklers must be installed in the piping chase, unless the chase is less than six inches in dimension.

ALTERNATE SUPPRESSION SYSTEMS (5.10)

♦   In certain applications, water may not be the most effective means to extinguish fires. In fact, in special cases, it could actually transport fire. Hence, the Uniform Fire Code allows alternate suppression systems.

♦   In some cases, the use of an alternative "suppression system" will not preclude the need to install water sprinklers.

ZONING (5.11)

♦   Zoning of fire sprinkler systems should be arranged so that, when activated, the emergency signal sent to the life safety alarm and monitoring system can pinpoint the area of the fire.

- General Fab Area,
- Interstitial Space or Plenum,
- Mechanical Support Space,
- Below Raised Floor,
- Solvent Exhaust Ducts,
- Acid Exhaust Ducts, and
- HPM Storage Rooms.

♦   Flow and tamper switches must be installed on each zone.

6

Electrical

6.1   ELECTRICAL POWER SYSTEMS

There are three classifications of backup systems to the normal electrical power system in H-2 storage and H-6 fab occupancies:

1.   Emergency Power Systems.
2.   Legally-required Standby Power System.
3.   Optional Standby Power System.

6.1.1   Emergency Power Systems:

Emergency power is required for facility systems which are considered essential for life safety. These systems take priority over standby systems where they share an alternate source of supply or when load is shed. Upon loss of normal power, emergency power loads must be connected to the alternate power source within 10 seconds instead of the 60 seconds allowed for legally required standby loads.   *(Refer to* **Figure 6.1** *for illustration of a large engine-generator set.)* Wiring for emergency systems is required to be kept "entirely independent of all other wiring and shall not enter the same raceway, cable, box or cabinet with other wiring." The following code excerpts address the need for emergency power systems:

Figure 6.1 EMERGENCY POWER ENGINE-GENERATOR SET. (Photo courtesy of Caterpillar)

UFC 51.106(c)3 (Workstations within Fabrication Areas) Ventilation Power and Controls - Emergency Power

"The exhaust ventilation system shall have an emergency source of power...   The emergency power is allowed to operate the exhaust system at not less than one-half fan speed when it is demonstrated that the level of exhaust will maintain a safe atmosphere."

All industrial occupancies shall have emergency lighting in accordance with NFPA Section 5-9.

UFC 80.303(a)7A-D (Toxic and Highly Toxic Compressed Gas) Indoor Storage - Emergency Power

"Emergency power shall be provided in lieu of standby power for:

A   Exhaust ventilation, including the power supply for treatment systems,

B   Gas Detection Systems,

C   Emergency Alarm Systems, and

D   Temperature Control Systems."

UBC 902(h) (Requirements for Group H Occupancies) Construction, Height and Allowable Area - Emergency Power

"An emergency power system shall be provided in Group H, Divisions 6 and 7 Occupancies. The emergency power system shall be designed and installed in accordance with the Electrical Code to automatically supply power to the exhaust ventilation system when the normal electrical supply system is interrupted. . . "

6.1.2   Sources of Emergency Power (NEC and NFPA 70):

In selecting an emergency source of power, consideration shall be given to the occupancy and the type of service to be rendered, whether of minimum duration, as for evacuation of an assembly area, or longer duration, as for supplying an alternative source of power and lighting due to indefinite period of power failure from trouble either inside or outside the building.

NEC 700-12 (Emergency Systems) Sources of Power - General Requirements

"Current supply shall be such that, in the event of failure, of the normal supply to, or within, the building or group of buildings concerned, emergency lighting, emergency power, or both will be available within the time required for the application but not to exceed 10 seconds. The supply system for emergency purposes, in addition to the normal services to the building and meeting the general requirements of this section, shall be permitted to comprise one or more of the types of systems described in (a) through (e) below. . .

(a) **Storage Battery.** Storage batteries . . . . maintain the total load for a period of 1½ hours minimum... An automatic battery charging means shall be provided.

(b) **Generator Set**
(1)   A generator set driven by a prime mover acceptable to the Authority Having Jurisdiction and sized in accordance with Section 700-5.   Means shall be provided for automatically starting the prime mover on failure of the normal service and for automatic transfer and operation of all required electrical circuits. . .

(2)   Where internal combustion engines are used as the prime mover, an on-site fuel supply shall be provided with an on-premise fuel supply sufficient for not less than 2 hours full-demand operation of the system.

(3)   Prime movers shall not be solely dependent upon a public utility gas system for their fuel supply or municipal water supply for their cooling systems.

Means shall be provided for automatically transferring from one fuel supply to another where dual fuel supplies are used.

> **Exception:** Where acceptable to the Authority Having Jurisdiction, the use of other than on-site fuels shall be permitted when there is a low probability of a simultaneous failure of both the off-site fuel delivery system and power from the outside electrical utility company.

(4)   Where a storage battery is used for control or signal power, or as the means of starting the prime mover, it shall be suitable for the purpose and shall be equipped with an automatic charging means independent of the generator set.

(5)   Generator sets which require more than 10 seconds to develop power are acceptable providing an auxiliary power supply will energize the emergency system until the generator can pick up the load.

(c) Uninterruptible Power Supplies. Uninterruptible power supplies used to provide power for emergency systems shall comply with the applicable provision of Section 700-12(a) and (b).

(d) Separate Service. Where acceptable to the Authority Having Jurisdiction as suitable for use as an emergency source, a second service shall be permitted. This service shall be in accordance with Article 230, with separate service drop or lateral, widely separated electrically and physically from the normal service to minimize the possibility of simultaneous interruption of supply.

(e)   Connection Ahead of Service Disconnecting Means. Where acceptable to the Authority Having Jurisdiction as suitable for use as an emergency source, connections ahead of, but not within, the main service disconnecting means shall be permitted. The emergency service shall be sufficiently separated from the normal main service disconnecting means to prevent simultaneous interruption of supply through an occurrence within the building or groups of buildings served.

(FPN): See Section 230-82 for equipment permitted on the supply side of a service disconnecting means."

Discussion:

Although other means are acceptable to the code, the most common source of emergency power is an engine/generator set with a local, on-site fuel supply. Typically, wafer fab occupancies have sufficient required emergency power loads to warrant diesel-engine-driven generators of substantial size (frequently 500 to 1,000 KW). While a second source of primary service may be acceptable to the Authority Having Jurisdiction, such a source should be carefully evaluated to ensure it can reasonably be relied upon given a major failure of the power company. Uninterruptible power supplies of the size necessary for a wafer fab would be prohibitively expensive. (*Refer to Section 6.2 for further discussion.*)

6.1.3   Legally Required Standby Power Systems:

"Standby" power systems are intended to provide electric power for control of health hazards and to aid in fire fighting or rescue operations (smoke or fume exhaust). Upon loss of normal power, legally required standby systems must be connected to the alternate power source within 60 seconds instead of the 10 second limitation for emergency power systems. Wiring for legally required systems may occupy the same raceways as other general wiring. Refer to NEC Article 701.

The following code excerpts address the need for legally required standby systems:

UBC 902(g) (Requirements for Group H Occupancies) Construction, Height and Allowable Area - Standby Power

"Standby power shall be provided in Group H, Divisions 1 and 2 Occupancies and in Group H, Division 3 Occupancies in which Class I, II or III organic peroxides are stored. The standby power system shall be designed and installed in accordance with the Electrical Code to automatically supply power to all electrical equipment required by the Fire Code when the normal electrical supply system is interrupted."

UBC 902(h) (Requirements for Group H Occupancies) Construction, Height and Allowable Area - Emergency Power

"An emergency power system shall be provided in Group H, Divisions 6 and 7 Occupancies. The emergency power system shall be designed and installed in accordance with the Electrical Code to automatically supply power to all required electrical equipment when the normal electrical supply system is interrupted.

The exhaust system may be designed to operate at not less than one-half the normal fan speed on the emergency power system when it is demonstrated that the level of exhaust will maintain a safe atmosphere."

UFC 80.301(s) (Hazardous Materials) Storage - Standby Power

"When mechanical ventilation, treatment systems, temperature control, alarm, detection or other electrically operated systems are required, such systems shall be connected to a secondary source of power to automatically supply electrical power in the event of loss of power from the primary source. See the Electrical Code."

UFC 80.307(a)9 (Organic Peroxides) Indoor Storage - Standby power

"Standby power is not required for the storage of Class III, IV or V organic peroxides."

UFC 80.401(l) (Dispensing, Use & Handling) Standby and Emergency Power

"When mechanical ventilation, treatment systems, temperature control, manual alarm, detection or other electrically operated systems are required by other provisions of this division, such systems shall be connected to a standby source of power to automatically supply electrical power in the event of loss of power from the primary source. See the Electrical Code.

When highly toxic compressed gases or highly toxic, highly volatile liquids are used or dispensed, emergency power shall be provided in lieu of standby power on all required systems. See the Electrical Code."

6.1.4   Sources of Legally Required Standby Power (from NEC):

NEC 701-11 Sources of Power - Legally Required Standby Systems

"Current supply shall be such that in event of failure of the normal supply to, or within, the building or group of buildings concerned, legally required standby power will be available within the time required for the application but not to exceed 60 seconds.   The supply system for legally required standby purposes, in addition to the normal services to the building, shall be permitted to comprise one or more of the types of systems described in (a) through (e) below. Unit equipment in accordance with Section 701-11(f) shall satisfy the applicable requirements of this article..."

NEC 701-11 (Emergency Systems) Sources of Power - Legally Required Standby Systems

". . . . In selecting a legally required standby source of power, consideration shall be given to the type of service to be rendered whether of short-time duration or long duration . . . . "

(a) **Storage Battery.** . . .

(b) **Generator Set.** . . .

(c) **Uninterruptible Power Supplies.** . . .

(d) **Separate Service.** . . .

(e) **Connection Ahead of Service Disconnecting Means.** . . .

The requirements for each of the alternate power sources are essentially as outlined in NEC 700-12 referenced above for an emergency power source.

Discussion:

The provision allowing standby loads to be "off-line" for up to 60 seconds gives the designer and owner/operator an opportunity to integrate the standby power source with the emergency power source. Generally, the size of an emergency generator is dictated by the starting current requirements of large fan motors for the fab ventilation systems. Once started and running normally, the fans will draw substantially less current than the generator is capable of producing. With the 60-second start delay, the emergency generator may be able to safely handle additional loads of standby systems without the need to oversize the unit. The means of coordinating the start-up of all loads and investigating the capacity of the generator must be carefully considered. Engine-generator manufacturers typically have computer software available to analyze starting and operating capacities of emergency systems with a variety of types of loads. These manufacturers must be consulted before designing such a combined load system.

6.1.5   Optional Standby Power Systems:

These are for systems whose failure could cause such effect as damage
to process equipment, interruption of manufacturing processes, loss of
air flow to maintain cleanliness conditions, etc. There is no time
requirement for the resumption of power to optional standby systems
because, as the title implies, these loads are considered for a standby
power system at the discretion of the owner/operator.

Although not required by code, process engineers may want selected
production equipment on optional standby power. For example, if
nitrogen and oxygen purifiers cease to function, the high-purity gas
piping may become contaminated causing problems when power is
restored. If process cooling water stops flowing through expensive
equipment, will damage result? Loads which are not related to life
safety, yet are desired to be connected to the emergency power system
must be separated and controlled by their own transfer switches. This
will allow delaying connection of these loads to the generator until it
is up to rated speed and voltage after the life safety loads have been
connected. This will also allow the non-life-safety loads to be "shed,"
if necessary, due to a generator malfunction.

6.2   DESIGN CONSIDERATIONS

6.2.1   When designing emergency/standby systems for wafer fabrication
facilities, the load requirements generally limit the alternate power
source choices. Storage batteries and uninterruptible power supply
(UPS) systems are not practical for most facilities.

6.2.2   The code provides that, "where acceptable to the Authority Having
Jurisdiction," the alternate power source can come from a separate
service or ahead of the service disconnecting means. This would
allow connection to the normal power source in an adjoining building.
Many facilities now follow this practice to provide a redundant source
of power for selected loads when the normal power source is down for
maintenance. However, it is doubtful that the Authority Having
Jurisdiction would accept a second connection to the normal electric

utility due to the extent of potential hazards in a wafer fab. The Authority may accept two services, if they are widely separated electrically and physically, have separate transformers, and are supplied from separate utility substations, such that the likelihood of a simultaneous failure of both sources is <u>extremely</u> remote.

Only exhaust is required to be on emergency power; however, the exhaust must be balanced with make-up air, and, thus, some supply fans may also require emergency power. Running the exhaust system alone would result in extreme negative pressure within the fab, which could, in turn, result in damage to ducts and partitions. If the exhaust fans were to be operated without the make-up air fans, it is questionable how much air flow could be produced (the cleanroom cube is well sealed to prevent air infiltration); therefore, it is questionable whether or not a safe rate of ventilation could be maintained. For this reason, <u>it is our recommendation that make-up air handling be operated on emergency power in conjunction with the exhaust system.</u>

In order to maintain pressure hierarchies, it may be desirable to consider connecting a few HEPA recirculating fans to the emergency power. (*Refer to Chapter 4, Mechanical Heating, Ventilating and Air Conditioning, for further discussion of these considerations.*)

6.2.3   The alternate power source is permitted to supply emergency, legally required standby, and optional standby system loads where automatic selective load pickup and load shedding is provided as needed to assure adequate power to (1) the emergency circuits; (2) the legally required standby circuits; and (3) the optional standby circuits, in that order of priority.

The alternate power source is permitted to be used for peak load shaving, providing the above conditions are met; however, most facility operators are concerned that utilizing their emergency power supply as a load shaving device will compromise its reliability in the event of an emergency power outage.

6.2.4   Typically, a generator set driven by an internal combustion engine is utilized. In some cases, a steam turbine may be utilized to drive the generator. A storage battery used to start the prime mover is required to be provided with an automatic battery charging means. An on-site fuel supply sufficient to operate internal combustion engines at full load for a minimum of two hours is required to be available. Due to requirements for periodic testing, a larger fuel supply should be considered.

The Authority Having Jurisdiction may approve off-site fuel supplies for the engine, where experience has demonstrated their reliability; however, this should only be considered if there is a compelling reason not to utilize an on-site fuel source.

6.2.5   Most H-6 fabs have solvent exhaust systems, separate acid exhaust systems, make-up air systems, and recirculating air systems. The proper operation of a fab pressure central system dictates that these systems be interlocked. Pressure hierarchies must be maintained to keep from contaminating adjacent work zones and to maintain cleanliness by precluding particle infiltration. It would be very costly to connect all of the HVAC fans to the emergency power system. The code (UFC 51.106) allows the air flow to be reduced to half fan speed if the minimum required air flow to maintain a safe environment is still met. A sufficient quantity of fans can be designated to run on emergency power to meet the one-half flow requirement. Transfer switches and damper controls can be provided to connect the designated fans to emergency power. Before attempting to limit the emergency power system loads by selecting which of the HVAC fans will be connected to emergency power, a detailed evaluation of the ventilation system and potential hazards must be conducted by the mechanical engineer. The electrical and mechanical engineers must work carefully together to develop a fail-safe operating scenario.

6.2.6   The National Electric Code Article 700-12(b)5 requires the generator be sized to energize the loads connected to the "emergency system" within 10 seconds.   UFC Article 80 references the NEC as the governing code for design of the emergency power systems.   The interpretation is that the loads designated as "emergency power" must be connected to the emergency power system within 10 seconds.   The ten second requirement does not allow the emergency loads to be grouped and started in steps separated by a few seconds; therefore, the starting current of motors becomes the governing factor when selecting an engine generator.

6.2.7   The emergency generator has to be sized to handle the starting current of the motors connected to the emergency power system.   An electric motor started "across the line" will draw approximately six times the name plate current.   Reduced voltage starters will significantly reduce this starting current.   Auto transformer reduced-voltage-type motor starters are more reliable than solid state types and should, therefore, be considered for life-safety loads.   The formula for determining starting current for an autotransformer starter is given below:

$$I_{START} = I_{LOCKED\ ROTOR} * (\%TAP)^2 + 1/4 * I_{FULL\ LOAD}$$

For a 50-horsepower motor, the "across the line" starting current would be 390 amps.   Utilizing an autotransformer starter set at 50 percent taps, the starting current would be 114 amps. Autotransformer starters are large, requiring a large motor control center and floor space.   This may be costly, but it may be less costly than installing a larger emergency generator and associated switchgear.

6.2.8   The emergency generator set requires an on-site fuel supply. Typically, diesel is the preferred fuel.   Some facilities' environmental quality engineers may direct that diesel fuel tanks be installed in underground concrete (or other leakproof) vaults.   Vaults should be provided with combustible gas sensors,   adequate ventilation and receptacle and lights on emergency power.   In other locations, a double wall fiberglass tank may be acceptable.   The fuel tank should be sized based on the following:

1.   Shelf life of diesel fuel is about six months.
2.   The generator set is required to be tested weekly.
3.   There should be sufficient fuel for an extended outage.

6.2.9   Emergency generator systems should be provided with alarm panels to monitor engine malfunctions.   If the generators are in a remote, unattended location,   a remote alarm should be provided in a continuously monitored location.   Fuel tank low-level alarms should also be sent to the same location.

6.2.10   Transfer Switches:

Both automatic and manual transfer switches should be considered to allow switching power sources to loads or to allow a selection of main or standby motors.   In general, automatic transfer switches are provided for code-required emergency loads,   and manual transfer switches are provided to change from a main to a standby fan or pump,   or to change from a normal to redundant power source. Transfer switches for motor loads must be rated to handle motor locked rotor current.

An automatic transfer and bypass-isolation switch should be provided to manually permit electrical bypass and isolation of the automatic transfer switch that could not otherwise be tested and maintained without interrupting the load. Bypass of the load to either the normal or emergency power source with complete isolation of the automatic transfer switch should be   possible regardless of the status of the automatic transfer switch.   A bypass-isolation switch is capable of bypassing the load to either source.

If a momentary power "blip" caused by switching transfer switches during testing will cause problems in the fab, then a "make before break" type of transfer switch should be utilized.

6.2.11   The installation, testing, and maintenance of the emergency power system should comply with NFPA 110 "Emergency and Standby Power Systems."

(Refer to **Figure 6.2** *for a one-line schematic diagram of the relationship between emergency generator set and the normal and redundant power supplies for a facility.)*

6.3   EMERGENCY SHUT OFFS

Emergency shut offs are required for some systems in the fab area and HPM storage rooms.   Location and extent of control should be discussed with the local Authority Having Jurisdiction.

6.3.1   Fabrication Area:

According to the Uniform Fire Code, a manual control switch must be provided that, when activated, will shut down exhaust ventilation and turn off all lights in the fab (H-6) area except those required to sufficiently illuminate means of egress.

> **UFC 51.105(d)1   (Semiconductor Fabrication Facilities Using HPM) Fabrication Areas - Ventilation Requirements**
> "Exhaust ventilation shall be provided to produce not less than one cubic foot per minute, per square foot, floor area, and shall comply with Section 79.805(e)."

> **UFC 79.805(e)   (Flammable and Combustible Liquids) Use, Dispensing and Mixing Rooms - Ventilation**
> "Continuous mechanical ventilation shall be provided at a rate of not less than 1 cubic foot per minute, per square foot of floor area over the design area.   For ventilation system design, see the Building and Mechanical codes."

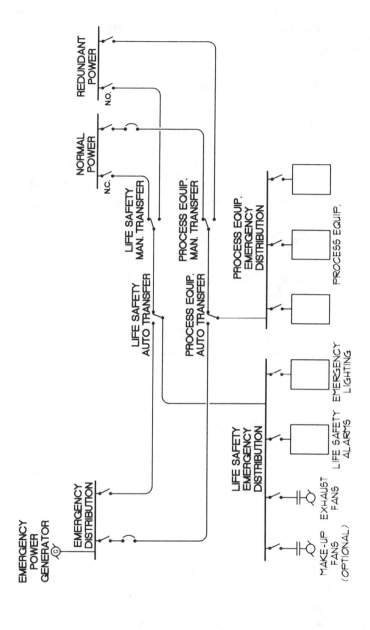

Figure 6.2 POWER DISTRIBUTION SCHEMATIC Illustrating Relationship between Normal Power, Redundant Power Supply and Emergency Power Source.

Another manual control switch is required for shutting down the recirculating fans (which supply air to the HEPA filter ceiling of the cleanroom).

UBC 911(b)3 (Division 6 Occupancies) Fabrication Area - Ventilation

"Mechanical ventilation. . . . Ventilation systems shall comply with the Mechanical Code except that the automatic shut-offs need not be installed on air moving equipment. However, smoke detectors shall be installed in the circulating airstream and shall initiate the signal at the emergency control station.

Except for exhaust systems, at least one manually operated remote control switch that will shut down the fab area ventilation system shall be installed at an approved location outside the fabrication area."

UFC 51.105(d)3 (Fabrication Area) Ventilation Requirements - Ventilation Controls

"There shall be a manual control switch for the supply or recirculation air systems, or both, located outside the fabrication area. The chief is authorized to require additional manual control switch locations."

Implementation of these emergency shut-off requirements requires coordination of various systems within the fab occupancy and possibly adjacent spaces.

We recommend that one switch be provided for the make-up air and exhaust fans, with a separate switch for the recirculating HEPA fans. The Uniform Fire Code requires that the lighting be shut-off if the exhaust is shut-off. The reasoning for interlocking the lighting system is that if there is no exhaust, personnel should not be in the fab. If the lights are off, it is reasoned that personnel will not enter the fab. In practice, the ventilation systems will not be shut down until the fab has been evacuated. An announcement can be made over the voice alarm system warning that the ventilation and lighting will be shut-off.

In some existing plants, exhaust fans may serve more than one fab (although this is no longer an accepted practice). If so, a damper system will be required to isolate one fab from another if they are to be shut down independently. As one fab is shut down, the exhaust requirement will decrease, requiring a change in the size or quantity of exhaust fans running. The control sequence for determining what fans should be running and what dampers should be closed can get complicated. It is recommended that, when a situation such as this is encountered, the two exhaust systems be separated, with discrete fans provided for each fab.

6.3.2   HPM Storage Rooms:

HPM storage rooms (H-2 Occupancies) must have smoke detection with automatic shutdown devices on air-moving systems over 2000 cfm. They are not exempt from this requirement as is the H-6 occupancy. HPM storage rooms must conform to UFC 79.804(3) (*see excerpt above*) by providing a manual control station to shut down the ventilation equipment and lights as described for the fab area above. A second manual control station or shut-off valve unique to HPM storage rooms is required.

> **UBC 905(b) (Requirements for Group H Occupancies) Light, Ventilation and Sanitation - Ventilation in Hazardous Locations**
> ". . . A manual shut-off control for ventilation equipment required by this subsection shall be provided outside the room adjacent to the principal access door to the room. The switch shall be of the break-glass type and shall be labeled 'Ventilation System Emergency Shut-off'."

UFC 51.110(a)2 (Storage of HPM) Outside Storage - Shut-offs in Piping

". . . A manual emergency shut-off valve located outside the building shall be installed on each HPM supply pipe from outside storage. The valve shall be identified, readily accessible and its location clearly visible."

UFC 80.401(c)2C (Dispensing, Use and Handling) Piping, Tubing, Valves and Fittings - Design and Construction

"Piping, valves, fittings and related components used for hazardous materials shall be in accordance with the following:

. . . Emergency shut-off valves shall be identified, and the location shall be clearly visible and indicated by means of a sign."

UFC 80.401(c)3D(i,ii) (Dispensing, Use and Handling) Piping, Tubing, Valves and Fittings - Supply Piping

"Readily-accessible manual or automatic remotely-activated fail-safe emergency shut-off valves shall be installed on supply piping and tubing at the following locations:

(i) the point of use, and

(ii) the tank, cylinder or bulk source."

UFC 80.301(m)4 Storage - Ventilation

"Exhaust ventilation systems shall comply with the following: . . .

A manual shut-off control shall be provided outside the room adjacent to the access door in the room or in a location approved by the chief. The switch shall be of the break-glass type and shall be labeled 'Ventilation System Emergency Shut-off'."

Since numerous HPM gases are piped into the fab area, it is usually best to provide a single manual station with all the HPM shut-off valves electrically interlocked in lieu of a separate valve for each HPM supply pipe. Electrical interlocking is relatively easy if done through the gas cabinet control panels. Before implementing such a design, however, the consequences of interlocking all gas sources should be discussed with the process engineers to determine if the risk to product is acceptable.

All emergency shut-off control stations should be mounted in some type of break-glass enclosure with the locations coordinated with the local fire chief. The code does not require that the switch has to be accessible to everyone; it could be located at the emergency control station or in a locked box adjacent to the fab, such as in an exit corridor. The shut-off switch need only be available to plant security personnel, management, or the fire service. There may be some concern that if a shutdown switch is readily accessible to the general population, a disgruntled employee or an uninformed employee could shut down the facility unnecessarily, resulting in many thousands of dollars of lost production. Again, the location must be approved by the enforcement official. The code also contains provisions for the fire chief to specify additional manual control switch locations.

Activation of any of the control stations will virtually halt production since, if the exhaust is shut down, all the process gases should automatically shut down as well. If the re-circulating HEPA fans are de-energized, cleanroom particle counts will rise rapidly, potentially ruining any in-process batches or exposed wafers. As you can see, these emergency shut-offs must be carefully engineered and located.

6.4   HAZARDOUS AREA ELECTRICAL REQUIREMENTS

The codes require the electrical installation, in certain areas, to meet explosion-proof requirements.

6.4.1   Work Station Environments (Fabrication Area):

UFC 51.106(e)   (Semiconductor Fabrication Facilities Using HPM) Workstations within Fabrication Area - Electrical Equipment

"Electrical equipment and devices within five feet of workstations in which flammable liquids or gases are used shall be in accordance with the Electrical Code for Class I, Division 2 hazardous locations.   Workstations shall not be energized without adequate exhaust ventilation.

> **Exception:** The requirements for Class I, Division 2 locations shall not apply when the air removal from the work station or dilution will provide nonflammable atmospheres on a continuous basis."

UBC 911(b)5 (Division 6 Occupancies) Fabrication Area - Electrical

"Electrical equipment and devices within the fabrication area shall comply with the Electrical Code. The requirements for hazardous locations need not be applied when the average air change is at least four times that set forth in Section 911(b)3 and when the number of air changes at any location is not less than three times that required by Section 911(b)3 and the Fire Code."

4 CFM SQ.FT.

The volumes of air moving through and exhausted from a wafer fab cleanroom typically meet the requirements of these exceptions. The extent of proof that the local building safety group requires will vary from jurisdiction to jurisdiction. Other sections of the codes require some exhaust to be on emergency power. Workstations handling flammable liquids must be included in the emergency exhaust system. If the air volume is sufficient, continuous, and on emergency power, the electrical installation within five feet of the workstation will not have to meet the requirements of Class I, Division 2 locations.

The UBC requirement of an air change rate of four times that set forth in 911(b)3 (thus, a total rate of at least 4 cfm/sq. ft.) may be interpreted by some building officials as requiring a portion of the fab recirculation air handlers to be provided with emergency power, if the exhaust/make-up air flow rate is not equal to 4 cfm/sq. ft.

6.4.2   The Uniform Fire Code Section on Storage Of Hazardous Production Materials (for H-2 occupancies) states:

> **UFC 51.110(b)8   Storage of HPM within Buildings - Electrical Requirements**
> "Electrical wiring and equipment located in HPM storage rooms shall be approved for Class I, Division 1 hazardous locations, and shall be in accordance with the Electrical Code.
>
> > **Exception:** When separate storage rooms are used, the storage rooms without flammable liquids or gases need not be approved for Class I, Division 1 hazardous locations."

The exception implies that explosion-proof electrical apparatus is required only in rooms where a flammable atmosphere may occur. In HPM storage rooms where flammable gases or liquids are stored in the open, the Class I, Division 1 electrical requirement would definitely be applicable. In rooms where the flammables are stored in monitored, Z-purged and protected (sprinklered) gas cabinets, Class I, Division 1 requirements may not apply for general electrical apparatus in the storage room.

Specially constructed gas cabinets can create isolated sub-atmospheres for the stored gases. Within this sub-atmosphere, all the requirements for UFC Article 51.107(b) are met. In a properly designed gas cabinet, all flammable gases are contained within the cabinet where the electrical equipment is rated Class I, Division 2. By definition, in a Division 2 area, an ignitable mixture is not normally present, but may occur under abnormal operating conditions, such as failure of process equipment (per NFPA 497A, Section 2-1.3). Since the only way an ignitable mixture could occur is by failure of the equipment, a properly designed cylinder storage cabinet meets the definition of a Division 2 atmosphere.

To be classified as Division 1, ignitable mixtures would have to be normally present during operation. Properly exhausted gas cabinets dilute flammable gases with air to a point well below the lower explosive limit, and the air in the cabinet is immediately exhausted at the rate of approximately 200 cubic feet per minute to the outside of the building.

In addition, in a properly designed fab, the gas cabinet exhaust is continually monitored by the toxic gas leak detection system and also for loss of air flow. Should either a gas leak or loss of exhaust flow be detected, the HPM gas will automatically be shut-off. The gas cabinets are designed so that the flammable gas cylinder valve is closed; all flammable gas within the piping is purged out with an inert atmosphere prior to opening the cabinet and disconnecting the cylinder for replacement. The gas cylinder should also be equipped with a restrictive orifice such that, even under the worst case conditions of full cylinder pressure, the gas released into the cabinet cannot exceed the lower flammable limit while the exhaust system is operating.

Gas cabinets whose exterior components meet the electrical requirements of Class I, Division 1 are not available as a standard manufactured item. A custom cabinet meeting these requirements could be built at great cost. Our interpretation is that a properly designed HPM storage room with exhausted, purged and sprinklered gas cabinets complies with the intent of the exception of Article 51.110(b)8 providing an equivalent installation. An HPM gas storage room will have up to 60 air changes per hour of exhaust ventilation as a result of the high concentration of gas cabinets. Furthermore, the gas cabinets operate at negative pressure with respect to the room, not allowing flammable vapors to escape into the room. Approval for this equivalent method of code compliance must be granted by the local Authority Having Jurisdiction. The design of ventilation systems and electrical safety devices must be carefully coordinated between the electrical and mechanical engineers to ensure an inherently safe facility.

UFC 80.307(a)10 (Organic Peroxides) Indoor Storage - Electrical Wiring and Equipment

"In addition to Section 80.301(r), electrical wiring and equipment in storage areas for Class I or II organic peroxides shall comply with the requirements for Class I, Division 2 locations."

UFC 80.307(b)3 (Organic Peroxides) Exterior Storage - Electrical Wiring and Equipment

"In addition to Section 80.301(r), electrical wiring and equipment in exterior storage areas containing Class I, II or III organic peroxides shall comply with the requirements for Class I, Division 2 locations."

UFC 80.308(a)6 (Pyrophoric Storage) Indoor Storage - Electrical Wiring and Equipment

"In addition to Section 80.301(r), electrical wiring and equipment in storage areas for pyrophoric gases and liquids shall comply with the requirements for Class I, Division 2 locations."

(Note: Revisions proposed for the 1994 UFC delete the requirement for special explosion-proof electrical devices for pyrophoric storage rooms.   The rationale is that the pyrophoric material will burn spontaneously, therefore, the electrical devices are not relevant.)

The Uniform Fire Code is more restrictive than the NEC, which states:

NEC 500-2 Hazardous (Classified) Locations - Location and General Requirements

"Locations are classified depending on the properties of the flammable vapors, liquids or gases or combustible dusts or fibers which may be present and the likelihood that a flammable or combustible concentrations or quantity is present. Where pyrophoric materials are the only materials used or handled, these locations shall not be classified..."

Pyrophoric materials ignite spontaneously upon contact with air. The use of electrical equipment suitable for a hazardous location may not prevent ignition of the material. Although the UFC 80.308 reference is more specific and restrictive, we believe classified electrical equipment in pyrophoric rooms is unnecessary; however, a thorough dialogue and approval of the Authority Having Jurisdiction concerning the issue is mandatory.

UBC 911(f)2F   (Division 6 Occupancies) Piping and Tubing - Installations in Exit Corridors and above Other Occupancies

"Hazardous production materials shall not be located within exit corridors or above areas not classified as Group H, Division 6 Occupancies except as permitted by this subsection.

Hazardous production material piping and tubing may be installed within the space defined by the walls of exit corridors and the floor or roof above or in concealed spaces above other occupancies under the following conditions: . . .

Electrical Wiring and equipment located in the piping space shall be approved for Class I, Division 2 Hazardous Locations.

> **Exception:** Occasional transverse crossings of the corridors by supply piping which is enclosed within a ferrous pipe or tube for the width of the corridor need not comply with Items A through F. "

6.4.3 If the fab has a solvent exhaust system with a solvent abatement system, the area around the solvent abatement facility must be in compliance with NFPA Article 36. Figure 2 in Article 36 shows the extent of the Class I, Division 1 and Class I, Division 2 areas. In addition, all tanks, vessels, motors, pipes, conduit, grating, and building frames must be electrically bonded together to control static electricity. A lightning protection system is recommended.

NFPA 36 5-5.1 Solvent Extraction Plants - Electricity

"Electrical wiring and electrical equipment of the extraction process, outward 15 ft. into the restricted area and vertically at least 5 ft. above the highest vent, vessel, or equipment containing solvent shall be installed in accordance with the requirements for Class I, Group C or D, Division 1 locations. "

NFPA 36 5-5.2 Solvent Extraction Plants - Electricity

"Electrical wiring and electrical equipment within the restricted area beyond the 15 ft. distance and to a height of 8 ft. above the extraction process grade level shall be installed in accordance with the requirements of Class I, Group C or D, Division 2 locations. "

NFPA 36 5-5.3 Solvent Extraction Plants - Electricity

"Electrical wiring and electrical equipment within the controlled area and within 4 ft. of the extraction process grade level, except the preparation process shall be installed in accordance with the requirements of Class I, Group C or D, Division 2 locations. "

SUMMARY

ELECTRICAL (6.0)

ELECTRICAL POWER SYSTEMS (6.1)

◆   There are three classifications of backup systems to the normal electrical power system in hazardous occupancies:

1.   Emergency Power Systems.
2.   Legally-required Standby Power System.
3.   Optional Standby Power System.

◆   Emergency Power Systems: Emergency power is required for facility systems which are considered essential for life safety. Upon loss of normal power, emergency power loads must be connected to the alternate power source within 10 seconds instead of the 60 seconds allowed for legally required standby loads. Wiring for emergency systems is required to be kept "entirely independent of all other wiring and shall not enter the same raceway.

■   **UFC 51.106(c)3 - Emergency Power**
"The exhaust ventilation system shall have an emergency source of power..."

■   All industrial occupancies shall have emergency lighting in accordance with NFPA Section 5-9.

■   **UFC 80.303(a)7A-D (Toxic and Highly Toxic Compressed Gas)**
"Emergency power shall be provided in lieu of standby power for:
A   Exhaust ventilation, including the power supply for treatment systems,
B   Gas Detection Systems,
C   Emergency Alarm Systems, and
D   Temperature Control Systems."

- **UBC 902(h) - Emergency Power**

    "An emergency power system shall be provided in Group H, Divisions 6 and 7 Occupancies to automatically supply power to the exhaust ventilation system when the normal electrical supply system is interrupted. . . "

◆ Legally Required Standby Power Systems: "Standby" power systems are intended to provide electric power for control of health hazards and to aid in fire fighting or rescue operations. Upon loss of normal power, legally required standby systems must be connected to the alternate power source within 60 seconds.

- **UBC 902(g) - Standby Power**

    "Standby power shall be provided in Group H, Divisions 1 and 2 Occupancies and in Group H, Division 3 Occupancies in which Class I, II or III organic peroxides are stored."

- **UFC 80.301(s) (Hazardous Materials) Storage - Standby Power**

    "When mechanical ventilation, treatment systems, temperature control, alarm, detection or other electrically operated systems are required, such systems shall be connected to a secondary source of power."

◆ Optional Standby Power Systems: These are for systems whose failure could cause such effect as damage to process equipment, interruption of manufacturing processes, loss of air flow to maintain cleanliness conditions, etc.

- Although not required by code, process engineers may want selected production equipment on optional standby power.

- Loads which are not related to life safety, yet are desired to be connected to the emergency power system must be separated and controlled by their own transfer switches.

■   The non-life-safety loads must be able to be "shed," if necessary, due to a generator malfunction.

♦   Only exhaust is required to be on emergency power; however, the exhaust must be balanced with make-up air, and, thus, some supply fans may also require emergency power. (*Refer to Chapter 5 for further discussion*).

♦   Generally, the size of an emergency generator is dictated by the starting current requirements of large fan motors for the fab ventilation systems. Once started and running normally, the fans will draw substantially less current than during across-the-line starting. With the 60-second start delay acceptable for standby power loads, the emergency generator may be able to safely handle additional loads of standby systems without the need to oversize the unit.

♦   The emergency generator set requires an on-site fuel supply.

♦   Transfer Switches: In general, automatic transfer switches are provided for code-required emergency loads, and manual transfer switches are provided to change from a main to a standby fan or pump, or to change from a normal to redundant power source.

EMERGENCY SHUT-OFFS (6.3)

■   Fabrication Area: A manual control switch must be provided that, when activated, will shut down exhaust ventilation and turn off all lights in the fab (H-6) area except those required to sufficiently illuminate means of egress.

■   Another manual control switch is required for shutting down the recirculating fans (which supply air to the HEPA filter ceiling of the cleanroom).

- "Except for exhaust systems, at least one manually operated remote control switch that will shut down the fab area ventilation system shall be installed at an approved location outside the fabrication area."

- We recommend that one switch be provided for the make-up air and exhaust fans, with a separate switch for the recirculating HEPA fans. The Uniform Fire Code requires that the lighting be shut-off if the exhaust is shut-off.

- HPM Storage Rooms: A manual control station or shut-off valve unique to HPM storage rooms is required.

- All emergency shut-off control stations should be mounted in some type of break-glass enclosure with the locations coordinated with the local fire chief. It could be located at the emergency control station.

HAZARDOUS AREA ELECTRICAL REQUIREMENTS (6.4)

- Work Station Environments (Fabrication Area):

    UFC 51.106(e)
    "Electrical equipment and devices within five feet of workstations in which flammable liquids or gases are used shall be in accordance with the Electrical Code for Class I, Division 2 hazardous locations.

    > **Exception:** The requirements for Class I, Division 2 locations shall not apply when the air removal from the work station or dilution will provide nonflammable atmospheres on a continuous basis."

    UBC 911(b)5
    "The requirements for hazardous locations need not be applied when the average air change is at least four times that set forth in Section 911(b)3."

-    The volumes of air moving through and exhausted from a wafer fab cleanroom typically meet the requirements of these exceptions.

-    If the air volume is sufficient, continuous, and on emergency power, the electrical installation within five feet of the workstation will not have to meet the requirements of Class I, Division 2 locations.

■   HPM Storage Facilities:

-    **UFC 51.110(b)8   Storage of HPM within Buildings - Electrical Requirements**
     "Electrical wiring and equipment located in HPM storage rooms shall be approved for Class I, Division 1 hazardous locations, and shall be in accordance with the Electrical Code.

     Exception: When separate storage rooms are used, the storage rooms without flammable liquids or gases need not be approved for Class I, Division 1 hazardous locations."

-    Explosion-proof electrical apparatus is required only in rooms where a flammable atmosphere may occur.

-    In rooms where the flammables are stored in monitored, Z-purged and protected (sprinklered) gas cabinets, Class I, Division 1 requirements may not apply for general electrical apparatus in the storage room.

-    In a properly designed gas cabinet, all flammable gases are contained within the cabinet where the electrical equipment is rated Class I, Division 2.

- Gas cabinets whose exterior components meet the electrical requirements of Class I, Division 1 are not available as a standard manufactured item.

- **UFC 80.308(a)6 (Pyrophoric Storage)**
"In addition to Section 80.301(r), electrical wiring and equipment in storage areas for pyrophoric gases and liquids shall comply with the requirements for Class I, Division 2 locations."

- *(Note: Revisions proposed for the 1994 UFC delete the requirement for special explosion-proof electrical devices for pyrophoric storage rooms.)*

- **NEC 500-2 Hazardous (Classified) Locations**
"Where pyrophoric materials are the only materials used or handled, these locations shall not be classified..."

- If the fab has a solvent exhaust system with a solvent abatement system, the area around the solvent abatement facility must be in compliance with NFPA Article 36.

7

Life Safety Alarm and Monitoring Systems

7.1   INTRODUCTION

Today, numerous alarm and monitoring systems are required by codes. These include fire alarm, smoke detection, sprinkler system supervision, emergency (spill) alarm and a continuous toxic gas monitoring and detection system. (*Refer to* **Figure 7.1** *for a basic "bubble diagram" of the various life safety systems to be integrated into a hazardous occupancy.*)

The UBC/UFC/NEC "H-6 Code Family" requires extensive life safety alarm and monitoring systems. We will present the code-required alarm and monitoring systems for semiconductor fabrication facilities while discussing potential strategies for upgrading a B-2 facility to H-6.

7.2   EMERGENCY CONTROL STATION (ECS)

7.2.1 Definition:

"Emergency Control Station (ECS) is an approved location on the premises of Group H, Division 6 Occupancy where signals from emergency equipment are received and which is continually staffed by trained personnel." (UBC 406).

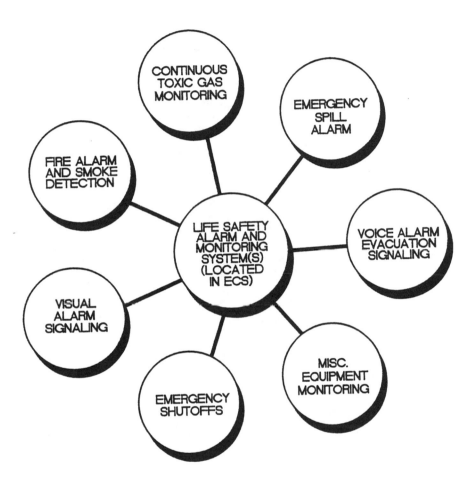

Figure 7.1 LIFE SAFETY ALARM INTERFACE Illustrating the Various Alarm and Monitoring Systems in a Hazardous Occupancy.

7.2.2   Code Requirement:

> **UFC 80.301(v)   (Hazardous Materials) Storage -**
> **Supervision**
> "When emergency alarm, detection or automatic
> fire-extinguishing systems are required in Sections 80.302
> through 80.315, such systems shall be supervised by an
> approved central, proprietary or remote station service or
> shall initiate an audible and visual signal at a constantly
> attended on-site location."

7.2.3   Location:

A prudent location for the ECS will be upstream of the prevailing
winds so that if an untreated catastrophic release of toxic gases
occurs, the emergency control station will remain safe for maintaining
critical life safety and control operations. Close coordination with the
local fire chief and enforcement authorities is requisite for approval.

The emergency control station should be located in a fire-resistant,
detached building or, as an alternate, located in the exterior wall of a
perimeter building that is easily accessible to emergency response
personnel. Separation must be maintained from the remaining portion
of the building by a two-hour wall (NFPA 72D-1.3). In addition, the
ECS must be constructed to conform with ANSI SE3.2 (UL 827)
which includes continuous monitoring by trained and competent
personnel, emergency lighting, and HVAC fed from an emergency
power source or generator.

The ECS can also house the proprietary system required under NFPA
Article 72D.

7.2.4   Design Consideration:

It may be possible to combine functions of the ECS with the plant's
main security station which is generally manned 24 hours per day.
An integrated console can be designed to incorporate fire alarm,

emergency spill alarm, toxic gas alarms, tank high level alarms, combustible gas alarms, emergency generator malfunction alarms, video CRTs with camera switches, card lock controls and monitoring, etc.

Systems of this magnitude must be user-friendly and should not require the operator to make critical decisions as to what course of action to take when an alarm is initiated. Ideally, the system will operate automatically while advising the operator what has occurred, relaying special instructions on a CRT. Means to record incidences for historical trending is recommended.

(Refer to **Figure** **7.2** *for a single-line diagram of an integrated life safety system. Note the relative locations of the various components.)*

7.3   ALARM AND MONITORING SYSTEMS

7.3.1   Fire Alarm, Smoke Detection and Sprinkler Supervision Initiating Devices:

The H-6 occupancy should be provided with either a manual or automatic supervised fire detection and alarm system installed and connected to the ECS in accordance with NFPA 72A. *(Refer to* **Figure** **7.3** *for a schematic illustration of the components of such a system.)*

A manual system is recommended since it is less expensive to install, reduces false alarms and since facilities of this nature are generally continuously monitored by trained supervisory personnel. False alarms in semiconductor manufacturing facilities must be minimized due to the extremely high production losses associated with evacuating the plant. The facility's insurance carrier should be consulted during design because they may have more stringent requirements than the governing codes.

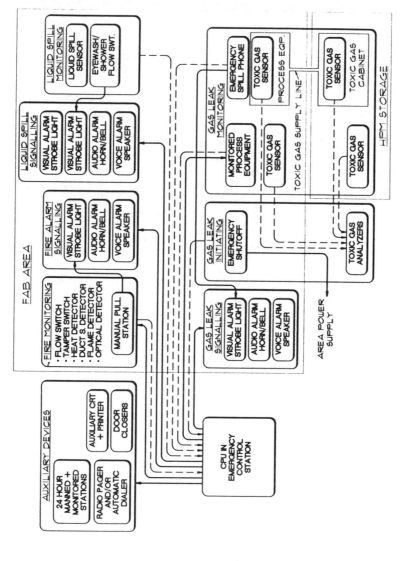

Figure 7.2 INTEGRATED LIFE SAFETY ALARM/MONITORING SCHEMATIC.

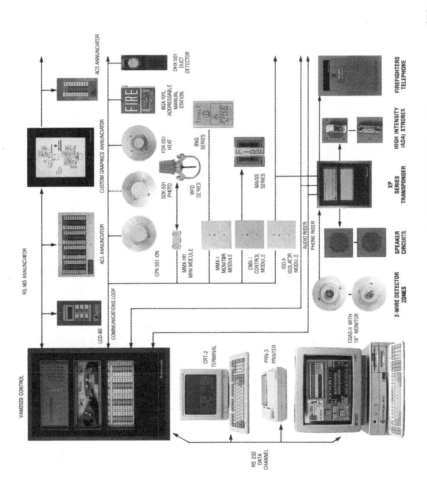

Figure 7.3 COMPONENTS OF INTEGRATED FIRE ALARM SYSTEM. (Photo courtesy of Notifier, Division of Pittway Corp.)

The code excerpts listed below reference fire alarm system requirements where hazardous materials are stored and for dispensing and use of hazardous materials.

UFC 80.402(b)3G(vi)   (Indoor Dispensing and Use) Special Requirements for Highly Toxic and Toxic Compressed Gases - Smoke Detection

"Smoke detection shall be provided in accordance with Section 80.303(a)10."

UFC 51-110(b)6   Storage of HPM within Buildings - Emergency Alarm

"A manual alarm box or approved emergency alarm-initiating device connected to a local alarm system shall be installed outside of each interior exit door which initiates a signal at the emergency control station and a local alarm when activated."

UFC 80.301(u)   (Hazardous Materials) Storage - Emergency Alarm

"An approved emergency alarm system shall be provided in buildings, rooms or areas used for the storage of hazardous materials. Emergency alarm signal device shall be installed outside of each interior exit door of storage buildings, rooms or areas. Activation of the emergency alarm-initiating device shall sound a local alarm to alert occupants of an emergency situation involving hazardous materials."

UFC 80.303(a)10   (Toxic and Highly Toxic Compressed Gases) Indoor Storage - Smoke Detection

"An approved supervised smoke-detection system shall be provided in rooms or areas where highly toxic compressed gases are stored indoors. Activation of the detection systems shall sound a local alarm."

UFC 80.306(a)10 (Liquid and Solid Oxidizers) Indoor Storage - Detection

"An approved supervised smoke-detection system shall be installed in all liquid and solid oxidizer storage areas. Activation of the detection systems shall sound a local alarm..."

UFC 80.307(a)12 (Organic Peroxides) Indoor Storage - Detection

"An approved, supervised smoke-detection system shall be provided in rooms or areas where Class I or II organic peroxides are stored, and where Class III or IV organic peroxides are stored in quantities exceeding the exempt amounts specified in Table No. 80.307-A. Activation of the detection system shall sound a local alarm.

> **Exception:** A smoke-detection system need not be provided in detached storage buildings provided with an automatic fire-extinguishing system."

UFC 80.401(m) Dispensing and Handling - Supervision

"Manual alarm, detection, and automatic fire-extinguishing systems required by other provisions of the division shall be supervised by an approved central, proprietary or remote station service or shall initiate an audible and visual signal at a constantly attended on-site location."

7.3.2 Types of Required Alarm Devices (*Refer to* **Figure 7.3**):

Manual Pull Stations

NFPA Article 72A requires the distribution of manual fire alarm pull stations throughout the protected area so that they are unobstructed, readily accessible and located in the normal path of exit from the area. Pull stations must be provided so that travel distance to the nearest station will not exceed 200 feet. Manual pull stations are also required outside of each HPM storage room's interior exit door (UFC 51-110-b6).

<u>Area Smoke and Heat Detectors</u>

Hazardous production material storage rooms are required to have a supervised smoke detection system in accordance with the above UFC requirements. In addition, NFPA 318-7 proposes:

NFPA 318-7 2-3.1

"An air sampling or particle counter-type smoke detection system shall be provided in the cleanroom return air stream to sample for smoke."

(Refer to **Figures** *7.4 and 7.5 for Illustrations of Air Sampling Smoke Detection Equipment and the Fire Growth vs. Response Time Characteristics of Air Sampling Systems.)*

Because the cleanroom circulates large quantities of air which tend to dilute smoke and other products of combustion, standard room detectors are not very effective. The air sampling smoke detection system draws air from various locations in the cleanroom on a continuous basis. This type of system responds to invisible pre-combustion gases that signify an incipient fire, which may not be recognized by a standard room smoke detector until the concentrations have reached a dangerous level. Thus, the air sampling system is capable of providing an early warning of an incipient fire condition.

The installation of heat detectors in all electrical rooms is an accepted practice.

- **NFPA 70 (Article 504)**

    Classifies an "Intrinsically Safe System" as an assembly of interconnected intrinsically safe apparatus, associated apparatus, and interconnecting cables in which those parts of the system which may be used in hazardous (classified) locations are intrinsically safe circuits. An intrinsically safe system may include more than one intrinsically safe circuit.

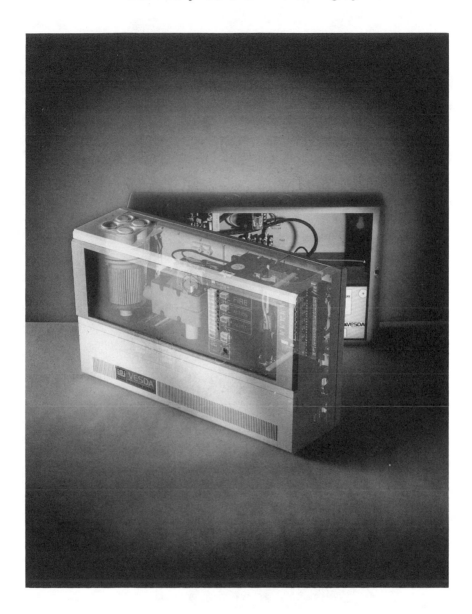

Figure 7.4   EARLY WARNING AIR SAMPLING TYPE SMOKE DETECTION SYSTEM.   (Photo courtesy of VESDA - Very Early Smoke Detection Apparatus)

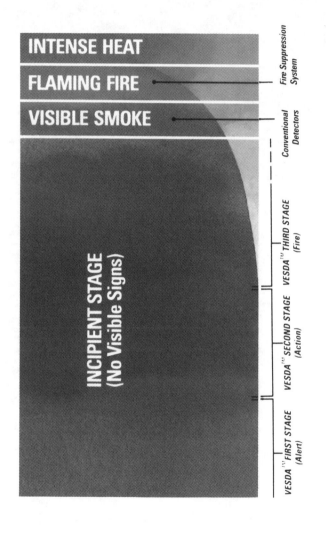

Figure 7.5 COMPARING AVERAGE RESPONSE of an Air Sampling Smoke Detection System to that of Conventional Detectors in an Escalating Fire Situation. (Photo courtesy of VESDA)

An "Intrinsically Safe Circuit" is a circuit in which any spark or thermal effect is incapable of causing ignition of a mixture of flammable or combustible material in air under prescribed test conditions. Test conditions are described in Standard for Safety, Intrinsically Safe Apparatus and Associated Apparatus for use in Class I, II, and III, Division 1, Hazardous (Classified) Locations, ANSI/UL 913-1988.

Duct Smoke Detectors

In all occupancies, duct-type smoke detectors are required in return air ducts for air handling units over 2,000 CFM (UMC 1009) and in the supply and return of units over 15,000 CFM (NFPA 90A). The purpose of these detectors is to stop the spread of smoke from one room to another by automatically shutting down air handler fans. A large fab can have over a hundred small air handlers. Automatically shutting them down one at a time as the smoke spreads through the room is ineffective in stopping the spread. The Uniform Building Code recognizes this and contains a provision unique to H-6 occupancies.

UBC 911(b)3 (Division 6 Occupancies) Fabrication Area - Ventilation

"...Ventilation systems shall comply with the Mechanical Code except that the automatic shut-offs need not be installed on air-moving equipment. However, smoke detectors shall be installed in the circulating airstream and shall initiate a signal at the emergency control station..."

Automatic shut-off to the motor need not be installed on air moving equipment when the smoke detector goes into alarm. However, a signal must be sent to the emergency control station. The code does require switches outside the fab for control of HVAC systems. Manual control switch(es) for the air handlers should be provided at the ECS where the smoke detectors annunciate. ECS personnel are charged with confirming/assessing the need for ventilating system(s) shutdown.

Smoke Door Holders

UBC Section 4306 defines the requirement for fire doors with mandatory self-closing provisions. Holders are required on doors through area separation walls (two-hour rating) and building separation walls (four-hour rating). Fire door locations are generally determined by the architect in conjunction with the code officials. Smoke detectors are required, on one or both sides of the door which, when initiated, release the door, allowing it to close automatically (NFPA 72E-9.5). These devices are intended to prevent the transmission of smoke from one space to another. Although not required, door holders can also be released on a command from the ECS.

Sprinkler Flow and Tamper Switches

Sprinkler flow and tamper switches to monitor the fire sprinkler systems are required by UBC-3802 and 3803. Typically, a plant is broken into many fire sprinkler zones. A single building may be broken into sub-buildings by fire separation walls with two- or four-hour ratings. It is recommended that each floor level exhaust duct system which is required to be sprinkled be monitored as a separate zone. The various fire sprinkler zones are each monitored separately to pinpoint the fire location and extent, minimizing response time by safety personnel.

Each zone includes a "flow switch" to detect the flow of water through the sprinkler riser and a "tamper switch" to advise the ECS that the water valve to that zone is not in the open position. Both the flow and tamper switches should be monitored at the ECS. Flow switches are Class "A" and tamper switches Class "B."

Optical Detectors

Use or dispensing of pyrophoric materials (materials which spontaneously react with the air) cannot be monitored by conventional smoke or heat detectors. Their detection speed is simply too slow. A much faster and more reliable "sensor" is suggested. Upon flame confirmation, the pyrophoric source must be immediately and automatically shut down. Optical sensors for each application require judicious selection.

Consider the pyrophoric silane (SiH_4). Silane reacts with oxygen in the air ($SiH_4 + 2O_2$) to produce, essentially, sand (SiO_2) and water (H_2O), while releasing energy in the form of a flame. Common infrared detectors on the market today are designed for hydrocarbon fires, looking only for infrared radiation in the CO_2 band. The silane reaction however, produces no CO_2 and, therefore, will not be detected. Ultraviolet detectors, on the other hand, monitor for -OH emissions (a component of water) which will detect a silane fire.

Adding to the complexity of sensor selection, it should be noted that some optical detectors are very susceptible to alarming from x-rays, lightning, arc welding and even radiant solar energy. Manufacturers are currently developing "smart" optical sensors that virtually eliminate false alarms without relinquishing fast response. However, there is a time lag between the time these sensors are developed and subsequently "listed" by Underwriters Laboratories. Most jurisdictions and insurance carriers require life-safety equipment listing by a recognized testing lab.

Hydrogen Sensors

In facilities utilizing gaseous hydrogen, it is recommended that sensitive, accurate, and reliable devices be used to provide detection and measurement of hydrogen gas for early warning and appropriate alarm initiation. Such devices include a hydrogen-sensing head as illustrated in **Figure 7.6**. The devices are designed to detect hydrogen leaks well before they reach hazardous concentrations.

The electrical classification of the transmitter assembly is required to comply with NEC Class 1, Division 1, Groups B, C, and D explosion-proof ratings. The assembly can also be operable as Intrinsically Safe for NEC Class 1, Division 1, Groups C and D when used with specified intrinsic safety barriers.

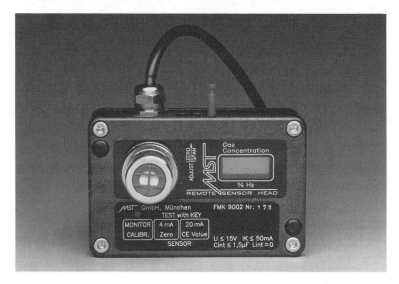

Figure 7.6 HYDROGEN SENSING HEAD. (Photo courtesy of MST Measurement Systems, Inc.)

7.3.3   Signaling and Annunciation:

(Signaling and annunciation functions for each fire alarm initiating device are shown in **Table 7.1 - Initiating Alarm Matrix.***)*

7.3.4   Upgrading B-2 Occupancies to H-6:

7.3.4.1 A first step in upgrading facilities from B-2 to H-6 occupancies is to construct the emergency control station. Since all monitoring and control circuits must be received and emanate from this station, the ECS must be in operation before any of the various alarm and monitoring systems can be properly operated. (See **Figure 7.1.**)

PROGRAMMED RESPONSE:	A RELEASE ALL DOOR HOLDERS ON COMM. LOOP	B INITIATE FIRE EVAC SIGNAL IN BUILDING (AUDIBLE MESSAGE #1)	C INITIATE FIRE EVACUATION SIGNAL IN ADJOINING BLDG. (AUDIBLE MESSAGE #1)	D INITIATE SPILL ALARM SIGNAL IN AFFECTED CORR. (AUDIBLE MESSAGE #2)	E INITIATE TOXIC GAS EVAC SIGNAL IN AIR HANDLING ZONE (AUDIBLE MESSAGE #3)	F ENERGIZE STROBE LIGHTS THROUGHOUT "BUILDING"	G ENERGIZE SPILL ALARM STROBE LIGHTS IN AFFECTED CORRIDOR	H SHUT-OFF VALVE OF DETECTED HPM	I SHUT-OFF VALVES OF ALL HPM IN HPM STORAGE RM	J NOTIFY ECS	K B2 OCCUPANCY SHUTDOWN AIR HANDLING UNIT IN "BUILDING"	L H-6 OCCUPANCY SHUTDOWN SUPPLY AIR SYSTEM IN "BLDG"	M SHUTDOWN EXHAUST IN "BUILDING"	N ALARM VERIFICATION PER TRANSPONDER
DOOR HOLDER SMOKE DETECTION	■									■				
INTERVENING DOOR BETWEEN "BUILDINGS DOOR HOLDER SMOKE DETECTOR	■									■				
DUCT SMOKE DETECTOR	■					■					■			
AREA SMOKE DETECTORS (B-2)	■										■			■
CIRCULATING AIRSTREAM SMOKE DETECTOR (H-6) (DUCT OR AREA TYPE)	■					■								
MANUAL PULL STATION	■													
MANUAL PULL STATION IN A CORRIDOR THAT SEPARATES TWO "BUILDINGS"	■					■								
FIRE SPRINKLER FLOW SWITCH	■									■				
FIRE SPRINKLER TAMPER SWITCH										■				
TOXIC GAS MONITOR LOW LEVEL ALARM										■				
TOXIC GAS MONITOR HIGH LEVEL ALARM	■				■		■			■		■		
SPILL ALARM PHONE				■			■			■				
HPM STORAGE ROOM LOCAL SHUT-OFF BUTTON									■	■				
WASTE STORAGE TANK HIGH LEVEL ALARM										■				
HEAT DETECTOR ALARM IN FLAMMABLE GAS CABINETS		■				■				■				
EMERGENCY GENERATOR MALFUNCTION ALARM										■				
FIRE PUMP RUNNING										■				

Table 7.1 INITIATING ALARM MATRIX.

(**Figure 7.2** is a typical flow diagram depicting the interface between the central processing unit (CPU) in the ECS and the alarm and monitoring subsystems.)

7.3.4.2 All manufacturing facilities require fire alarm and smoke detection systems. There are essentially no differences between the fire alarms in B-2 and H-6 occupancies with the exception of the duct smoke-detector requirements. In a B-2 occupancy, duct detectors must automatically shut down the fan upon detection of smoke, whereas in H-6 it does not. While in the process of upgrading the plant, it is recommended that automatic shut-offs be left active on all existing air handlers until individual conversions are completed. At that time, it is a relatively simple matter to remove automatic shut-off features from the smoke detectors. This strategy, however, must be approved by the local code enforcement officials.

7.4   CONTINUOUS TOXIC GAS MONITORING

7.4.1   A continuous gas-detection system is required by Uniform Fire Code Article 51 to monitor for the presence of toxic gases when the maximum safe level of long-term exposure could be reached. The system must continually monitor use areas for the presence of fugitive gases that might be present but unknown due to the lack of perception by building occupants. Continuous monitoring is also required for flammable gases or vapors. (*Refer to* **Figures 7.7 and 7.8** *for illustrations of two different types of toxic monitoring/alarm systems.*)

These systems must be designed in accordance with:

- UFC Article 51 and 80
- UBC Chapter 9
- Process Equipment Needs
- Plant Safety and Insurance Requirements

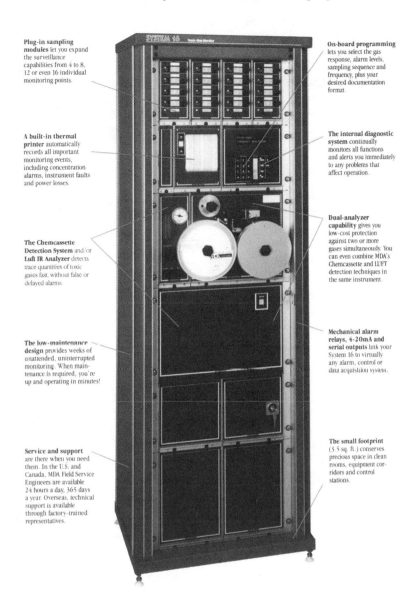

Plug-in sampling modules let you expand the surveillance capabilities from 4 to 8, 12 or even 16 individual monitoring points.

A built-in thermal printer automatically records all important monitoring events, including concentration alarms, instrument faults and power losses.

The Chemcassette Detection System and/or **Luft IR Analyzer** detects trace quantities of toxic gases fast, without false or delayed alarms.

The low-maintenance design provides weeks of unattended, uninterrupted monitoring. When maintenance is required, you're up and operating in minutes!

Service and support are there when you need them. In the U.S. and Canada, MDA Field Service Engineers are available 24 hours a day, 365 days a year. Overseas, technical support is available through factory-trained representatives.

On-board programming lets you select the gas response, alarm levels, sampling sequence and frequency, plus your desired documentation format.

The internal diagnostic system continually monitors all functions and alerts you immediately to any problems that affect operation.

Dual-analyzer capability gives you low-cost protection against two or more gases simultaneously. You can even combine MDA's Chemcassette and LUFT detection techniques in the same instrument.

Mechanical alarm relays, 4-20mA and serial outputs link your System 16 to virtually any alarm, control or data acquisition system.

The small footprint (3.5 sq. ft.) conserves precious space in clean rooms, equipment corridors and control stations.

Figure 7.7   TOXIC GAS MONITORING CABINET Utilizing Chemical Cassette Technology.   (Illustration courtesy of MDA Scientific)

Figure 7.8   TOXIC GAS MONITORING UNIT.   (Photo courtesy of TeloSense Corporation)

It is prudent to evaluate the hazardous material requirements of the Uniform Fire Code Article 80. In the 1988 edition, approximately 60 pages were amended to the 1985 Article, along with associated appendices. Article 80 implements general requirements for all types of hazardous materials regardless of the occupancy or use, whereas Article 51 and UBC 409 deal with specific materials utilized in semiconductor manufacturing. Article 51 and the UBC are generally more restrictive. In the event that process materials are not addressed in the applicable H-6 codes, then Article 80 will govern. Although some cities may not have adopted the 1991 UFC, verification should be obtained from the local Authorities Having Jurisdiction.

Protection must be provided where personnel may be exposed to gases that are either toxic or highly toxic. This philosophy is often difficult to implement because of the uncertainty of just how to define the word "may." Apprehensions among employees and the public surrounding toxic gas exposure are strong, and there has been a flurry of activity in both the technical and regulatory communities to address both the fears and perceived hazards.

7.4.2   The following code excerpts address the requirement for gas detection systems:

UFC 51.105(e)3 (Fabrication Areas) Special Provisions - Gas Detection Systems
"When hazardous production material gas is used or dispensed and the physiological warning properties for the gas are at a higher level than the accepted permissible exposure limit (PEL, as defined by OSHA 29CFR 1910.1000) for the gas, a continuous gas detection system shall be provided to detect the presence of a short-term hazard condition. . . When dispensing occurs and flammable gases or vapors may be present in quantities in excess of 20 percent of the lower explosive limit, a continuous gas detection system shall be provided. The detection system shall be connected to the emergency control station."

UFC 80.303(a)9 (Toxic and Highly Toxic Compressed Gases) Indoor Storage - Gas Detection

"A continuous gas-detection system shall be provided to detect the presence of gas at or below the permissible exposure limit or ceiling limit. The detection system shall initiate a local alarm and transmit a signal to a constantly attended control station (ECS). The alarm shall be both visual and audible and shall be designed to provide warning both inside and outside of the storage area. The audible alarm shall be <u>distinct</u> from all other alarms.

> **Exceptions:**
>
> 1   Signal transmission to a constantly attended control station need not be provided when not more than one cylinder is stored.
>
> 2   A continuous gas-detection system need not be provided for toxic gases when the physiological warning properties for the gas are at a level below the accepted permissible exposure limit for the gas.

The gas-detection system shall be capable of monitoring the room or area in which the gas is stored at or below the permissible exposure limit or ceiling limit and the discharge from the treatment system at or below one-half the IDLH limit."

7.4.3   Automatic shutdown of the gas cabinet supply lines, whether used for storage or dispensing, is required.   Continuous monitoring of flammable and toxic gases is also required under UFC Articles 51 and 80 and requires automatic shutdown for supply of toxic gases when utilized within the fab area, HPM storage rooms, piping over or in exit corridors, and gas cabinets. Upon detection and shutdown, an alarm must be initiated at the ECS.

The following code excerpts address the requirement for automatic shut-off valves.

UFC 51.107(b)2 (Storage and Dispensing of HPM within Fabrication Areas) Special Requirements for HPM Gases - Gas Detection

"Gas cabinets for HPM gases shall be provided with a continuous gas monitoring system in accordance with Section 51.105(e)3 regardless of whether dispensing occurs. Activation of the detection system shall automatically shut the valves on all HPM gas lines from the cabinets and initiate an alarm to the emergency control station."

UFC 80.402(b)3F(v) (Indoor Dispensing and Use) Spill Control, Drainage Control and Secondary Containment - Gas Detection

"Gas detection shall be provided in accordance with of Section 80.303(a)9. Activation of the monitoring system shall automatically close the shut-off valve on highly toxic or toxic-gas-supply lines related to the system being monitored ..."

UFC 80.402(c)8D (Exterior Dispensing and Use) Special Requirements for Highly Toxic or Toxic Compressed Gases - Gas Detection

"Gas detection shall be provided in gas cabinets and exhausted enclosures in accordance with Section 80.303(a)9. Activation of the monitoring system shall automatically close the shut-off valve on highly toxic or toxic gas supply lines related to the system being monitored. . ."

7.4.4   The definition for a Hazardous Production Material (HPM) is found in the following code excerpt from the UBC:

> **UBC 409 Definitions and Abbreviations - Hazardous Production Material (HPM)**
>
> "Hazardous production material (HPM) is a solid, liquid or gas that has a degree of hazard rating in health, flammability, or reactivity of 3 or 4 and which is used directly in research, laboratory, or production processes which have, as their end product, materials which are not hazardous."

The following is a partial listing of the gases that meet this criteria and should be monitored:

PH_3	-	Phosphine
AsH_3	-	Arsine
SiH_4	-	Silane
SiH_2Cl_2	-	Dichlorosilane (DCS)
B_2H_6	-	Diborane
CO	-	Carbon Monoxide
NF_3	-	Nitrogen Trifluoride
BF_3	-	Boron Trifluoride
BCl_3	-	Boron Trichloride
GeH_4	-	Germane
NH_3	-	Ammonia
CFl_4	-	Carbon Tetrafluoride
HCl	-	Hydrogen Chloride
H_2S	-	Hydrogen Sulfide
Cl_2	-	Chlorine

Although the above listing is not a complete list of gases that require monitoring, it is helpful as a guide. The quantities of gas being used and the manufacturing process also have a bearing on the monitoring requirements. The final decision as to which gases are to be monitored is up to Plant Safety and the local Authorities Having Jurisdiction.

(NOTE: Plant safety/risk management and insurance standards may be more strict than the codes! Confer with the responsible group for a ruling on all chemicals and gases.)

7.4.5   Standards for chronic exposure limits to employees are generally defined by the threshold limit value (TLV) concept (as calculated on a time weighted average [TWA] basis), promulgated by the American Conference of Governmental Industrial Hygienists. On the other hand, the limits of exposure to the general public for acute exposure are evolving.

7.4.6   Sensor Locations:

The codes do not specify the exact location of gas sensors inside the fab area; they simply state that the gas must be detected. Extremely high air movement rates play a major role on sensor locations and spacings. Because of this, the manufacturer of the selected detection systems should be consulted. It is advisable to locate some sensors around process equipment utilizing toxic gases in plain sight of the fab workers for service and psychological reasons. As stated earlier, additional sensors are required in the gas cabinets, HPM storage rooms, exit corridors (when toxic gas piping is present) and also in the exhaust stacks.

A typical scenario follows:

A cylinder of HPM gas is located inside an exhausted gas cabinet. A sensor is placed in the exhaust duct, a distance of five times the duct diameter downstream from the gas cabinet. A sensor is also placed in the room where the gas cabinet is located. The HPM gas piping passes over an exit corridor on its way to the fab. A fire rated chase must be constructed for the piping over the corridor. The process equipment that uses the HPM gas should have an exhausted enclosure for the area where the HPM gas is utilized. A sensor should be placed in the process equipment's exhaust duct a distance of five duct diameters downstream from the exhausted enclosure. Sometimes a sensor is placed in the process equipment adjacent to where the gas could possibly leak out.   Another sensor is placed in the general vicinity of the operator's "breathing zone."

7.4.7   Control, Signaling and Levels of Detection:

It is recommended that two levels of detection be provided, a low warning level set at 1/2 TLV (not code-required, but strongly recommended) and a high alarm level set equal to TLV (code-required).   Low level detection signal should be sent to the ECS so that trained personnel can evaluate and respond to the situation before it reaches dangerous levels. This enhances employee life safety and prevents production losses due to automatic shut downs.   If the high level is attained, an alarm is sent to the ECS, a distinct toxic gas evacuation signal must be broadcast locally throughout the area, and the detected gas supply line must be automatically shut off (UFC 80.303-a-9).   *Local area* in this sense refers to any area sharing a common air handling and distribution system (in some fab designs, this may be considered the entire fab).

7.4.8   Gas Analyzers:

Gas analyzers are manufactured by several different companies, Perkin Elmer, MDA Scientific and Telos Labs, to name but a few. The analyzers should be discriminatory whenever possible to avoid nuisance tripping.   Also the definition of "continuous monitoring" requires that each sensor or location be sampled and evaluated in periods not exceeding 30 minutes.   Gas analyzer outputs should contain digital signals, one for 1/2 TLV low-level warning, and one for TLV high-level alarm. These signals must interface between the gas analyzers and the life-safety alarm and monitoring system. Life-safety systems are required to be electrically supervised to provide indications of opens and shorts in wiring or malfunctioning components. The monitoring system that takes signals to the ECS and directs control of life-safety systems must meet this requirement. Most gas analyzers have a limited number of control contacts.   The use of standard relays and field-fabricated controls to handle gas shut-off, initiation of local alarms, etc. does not provide supervision. The signal from the gas analyzer should be taken to a U.L. "listed" monitoring system and the alarms and control sequences distributed from there.

Some gas detectors draw in air samples from the monitored space through 3/8 inch sample tubing. It is recommended that this tubing be installed in conduit or a wireway system for mechanical protection and servicing. Some gas analyzers will sense and alarm if the sensor tubing is completely blocked or severed. However, a small cut or pinhole that dilutes the sample could go undetected.

The time required to detect toxic gas is dependent upon three main criteria:

1.   Type of gas being monitored,
2.   Concentration of the gas, and
3.   Set level of detection

Generally the higher the concentration, the faster the response time, e.g., the response-analysis time for phosphine, utilizing MDA's System 16 or PSM 8XT analyzers, is as follows: 1/2 TLV (150 PPB) at 13 seconds and TLV (300 PPB) at 7 seconds. For arsine, it takes approximately 30 seconds to initiate an alarm at 1/2 TLV.   An on-board microcomputer determines the optimum analysis time length based on the specific gas and programmed alarm setting. This time cannot be field-modified; it is factory-set via software.

All gas analyzers should be located outside the fab. If an alarm is received at the ECS, personnel will need access to the analyzer to determine the extent of the problem. In addition, maintenance efforts will be difficult and disruptive if analyzers are located in cleanrooms.

Alternatively, there are discrete gas detectors manufactured that require placement in the detection area with electrical wiring in lieu of pneumatic tubing running back to the remote control panel. The drawback to these detectors is that testing and maintenance must be done "gowned up" which interferes with fab operations. All installations must be carefully tested in accordance with the manufacturer's specifications.

There is a tendency for the equipment engineers in different fabs to handle HPM gas monitoring differently. This lack of common standards becomes a real problem when discrete systems are tied to a common alarm and monitoring system. Monitoring methods and procedures should be consistent throughout the plant. Maintenance and operation of the entire system will be more manageable if the system is consistent.

One type of toxic gas monitor operates by exposing a chemically-treated tape, called a "chemcassette," to an air sample that is pulled through a teflon sampling tube from a designated location. Each chemcassette is specially formulated to react only with a particular gas or family of gases. When exposed to the target gas, it changes color in direct proportion to the amount of gas present. The higher the concentration, the darker the stain. Color changes are quantified with the electro-optical detection system by continually measuring the amount of light reflected off the chemcassette. (*Refer to* **Figure 7.9** *for an illustration of this sensing technique.*)

The sequence of monitoring is completely programmable. This feature is useful where one or two monitoring locations require significantly more or less surveillance than other monitoring points. The instrument will automatically provide balanced coverage if no custom sequence is chosen. It is recommended that high traffic areas be monitored more frequently than unmanned storage areas. However, every channel should be sampled at least once every 30 minutes. When a multiple channel device is chosen, it must be confirmed that all channels can be monitored within this 30-minute code criteria.

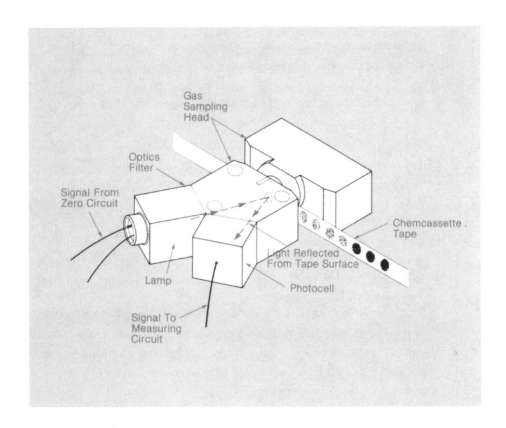

Figure 7.9 DETAIL OF TOXIC GAS MONITORING SYSTEM Utilizing Chemical Cassette Technology. (Illustration courtesy of MDA Scientific)

Each channel or point may have two detection levels that can be predetermined and programmed into the system. Each channel also has a set of contacts for high and low levels. For maximum protection of employees, both levels can be utilized even though only one, the TLV, is required by code. A warning level of 1/2 TLV (established by the facility safety department) is programmed into the monitoring device which will activate a set of contacts upon detection. If the leak continues and reaches TLV, a second set of contacts are activated. Thus, two levels of monitoring are provided.

To summarize, a typical monitoring system, such as the MDA System 16 Analyzer, has the following output relays:

1.   Self diagnostic contacts that close upon low flow, photocell malfunction, analyzer tape problem and printer failure.

2.   Low- and high-level contacts per channel (two discrete contacts per channel).

3.   Two master contacts for the complete monitoring of the entire console. One contact will close upon the closing of any of the 16 monitoring points warning contacts and one for any of the alarm contacts.

Each contact in the MDA is a single-pole double-throw with normally open, normally closed and common terminals. These contacts are generally monitored by the life-safety alarm and monitoring system and are sent to the ECS where control-by-event functions are activated and a custom program message printed.

7.4.9   Interfacing with the Life Safety Alarm and Monitoring System:

As stated previously, the Uniform Fire Code requires the following actions upon TLV detection:

1.   Notify the ECS,
2.   Automatically shut down the supply of the detected gas, and

3.   Automatically sound local alarm, evacuating the area.

Since analyzers typically have only a single set of high-level alarm contacts and numerous unique actions must occur, the life-safety alarm and monitoring system must be interfaced with the analyzers. Upon change due to TLV detection, the system must perform the items listed above.

7.4.10 Bypassing the Life Safety Alarm and Monitoring System:

Inevitably, someone asks, "Can the life-safety alarm and monitoring system be bypassed, by utilizing the analyzer contacts only?" Yes, it is possible, if properly arranged. However, there are several reasons why we strongly discourage this:

1.   In a fire, all supplies of toxic and reactive gases should be cut off, an impossibility if they were interlocked to the analyzers alone. If they are interlocked with the life safety alarm and monitoring system, all cabinets can be shut down by a single switch or a command at the ECS (or any remote location as desired).

2.   No maintenance could be performed on the analyzers without resulting in the gas cabinets being shut down due to loss of control circuit power. This would result in huge production losses any time routine maintenance had to be performed on the analyzers.

3.   Distinct annunciation pertaining to the type of toxic gas that has been detected cannot be relayed to the ECS. This critical information quickens response and clean up.

4.   Gas analyzers shutdown channel contacts may have to be wired in series with other contacts (possibly located in other analyzers), if more than one sensor monitors for release of a

gas fed from a common gas cabinet (which is usually the case). This may tend to propagate electromagnetic interference (EMI) resulting in false alarms.

In addition, utilizing the life safety alarm and monitoring system offers excellent flexibility for future gas panels, gas sensors and gas analyzer modifications while allowing for control-by-event functions.

It is, therefore, recommended that all code-required functions be handled through the life-safety alarm and monitoring system.

7.4.11   Power Requirements:

Connection to an emergency power source is requisite for all HPM gas analyzers. Typically this source is the plant's emergency generator system. Upon loss of normal power, the analyzer will be disabled for approximately ten seconds until the emergency generator comes up to speed. The gas analyzers will begin monitoring again the moment power is restored. If the particular analyzer utilized must go through a "boot-up" process each time it is powered up, a UPS (uninterruptible power system) should be considered.

7.4.12   B-2 to H-6 Upgrade Phasing:

Toxic gas monitoring systems should be immediately upgraded to conform to all the new applicable codes. Not only does it protect employees but it also reduces the facility's liability. Toxic gases are potentially the most dangerous aspect of fab operation since, in many cases, they are an invisible and odorless threat to life.

7.5   EMERGENCY SPILL ALARM

7.5.1   In new H-6 facilities, transporting HPMs in code-required exit corridors is prohibited. This is the reason for the development of service corridors which provide alternate paths for HPM delivery. In existing facilities, the Uniform Fire Code will allow transportation of HPMs in the exit corridors only if the following conditions are met:

1.   All containers are of an approved type.
2.   Quantity to be transported is restricted.
3.   An approved cart may be used to increase quantities transported.
4.   Spill-alarm signaling devices are provided at 150-foot intervals.
5.   Sprinkler system suitable for ordinary hazard Group 3.
6.   No storage or dispensing is permitted.
7.   HPM piping entering the corridor is to be provided with:
    a.   Fire resistant enclosure of one-hour construction.
    b.   Ventilation rate of six changes per hour.
    c.   Drainage receptacle.
    d.   Shut-off valves at points of entry/exit.
    e.   Excess flow control.
    f.   Gas detection.
    g.   Electrical requirements of Class I, Division 2 (areas within the piping chase).
    h.   All lines containing HPMs are identified in accordance with the standards.

7.5.2   The purpose of the HPM emergency spill alarm is to initiate a local alarm and to report the conditions of the spill, leak or other accident within the exit corridor used to transport HPMs.

The code excerpts listed below reference alarm requirements in exit corridors:

> **UFC 51.108(b)8   (Handling of HPM within Exit Corridors) Existing Buildings - Emergency Alarm**
> "An emergency telephone system, a local alarm manual pull station or an approved signaling device shall be provided at not more than 150-foot intervals or fraction thereof and at

each exit stair doorway. The alarm signal shall be relayed to the emergency control station, and a local signaling device shall be provided."

UFC 80.403(d) (Hazardous Materials) Handling - Emergency Alarm

"When hazardous materials rated 3 or 4 in accordance with UFC Standard No. 79-3 are transported through exit corridors or exit enclosures, there shall be an emergency telephone system, a local manual alarm station or an approved alarm-initiating device at not more than 150-foot intervals and at each exit doorway throughout the transport route. The signal shall be relayed to an approved central, proprietary or remote station service or constantly attended on-site location and shall also initiate a local audible alarm."

7.5.3 Although a manual pull station is acceptable by the code, it is recommended that utilizing a telephone mounted in an enclosure be considered. This allows workers to inform the ECS of the exact nature and severity of the spill, increasing the speed and efficiency of response personnel. All phones should have their own dedicated circuits and should be approved by the local Authority Having Jurisdiction. Removal of the phone from its cradle or activating the initiating device must automatically send a signal to notify the ECS and automatically broadcast a distinct spill alarm evacuation signal throughout the affected corridor. Electrically supervised "fireman's telephone" systems, developed for use in high-rise buildings, are easily adapted for this use. Where more than one spill-alarm device is located in a common corridor, we also recommend mounting a strobe light above, activated when the phone is removed from the cradle, to aid the emergency response personnel in locating the spill. In addition, a spill alarm-initiating device should also be located at each HPM storage room. Although not a code requirement, placement of an emergency spill alarm phone at bulk materials storage and handling areas is recommended.

7.5.4   Spill Containment Areas:

> **UFC 80.301(l)4   (Hazardous Materials) Storage - Spill Control, Drainage Control and Secondary Containment**
> ". . . When secondary containment is required, a monitoring method capable of detecting hazardous material leakage from the primary containment into the secondary containment shall be provided. Visual inspection of the primary containment is the preferred method; however, other means of monitoring are approved by "the chief". Where secondary containment may be subject to the intrusion of water, a monitoring method for such water shall be provided.   Whenever monitoring devices are provided, they shall be connected to distinct visual or audible alarms."

Another good practice for detecting chemical spills is to install detection systems in the bottom of piping trenches below the fab area's raised floor. If chemicals are detected, due to a ruptured line or spill, the ECS should be notified and, again, the spill alarm evacuation signal should be broadcast throughout the affected area.

7.5.5   B-2 to H-6 Upgrades:

Emergency spill stations must be installed in all exit corridors used to transport HPMs. Exit corridors by their very nature are designed to maintain a safe passage for employee evacuation. Hence, accidents in these corridors that place employees at risk when exiting must be immediately reported and a local alarm sounded to prevent workers from entering the contaminated area. Secondary means of egress must be provided and training given to employees explaining their use. Plant management must enforce the use of designated chemical transport corridors.

7.6   AUDIBLE ALARM EVACUATION

7.6.1   The voice alarm evacuation system consists of high-reliability speakers located throughout the facility to reduce confusion and help eliminate panic in an emergency evacuation of a given area.   The following distinct audible alarm systems are required:

1.   Fire alarm evacuation signal.
2.   Toxic gas evacuation signal.
3.   Chemical spill evacuation signal.

Audible broadcasting should utilize either tones, pre-recorded messages, integral paging or any combination thereof.   Pre-recorded messages have the advantage of broadcasting a calm tone along with a short descriptive message reducing the potential for employees to panic in emergency situations. Pre-recorded messages also eliminate the need for employees to remember what signal goes with each type of emergency.

7.6.2   Evacuation-speaker zoning should be carefully planned and laid out. For fire alarm systems, the area enclosed by fire-separation walls should be considered an annunciation zone.   For HPM gas monitoring/alarms, each area with its own air handling system could be considered as a zone.   Sub-zones may be further selected on the basis of particular processes being performed.

According to the NFPA, speakers designed for broadcasting of emergency life-safety alarms must maintain 15 decibels above the ambient noise level. Because the ambient noise level in a new facility must be estimated, and changes frequently occur due to the addition or removal of production equipment, it is recommended that the speakers provided contain multiple-tap power ratings. If the sound level in an area is too loud or too soft, the speaker taps can be adjusted, changing the volume accordingly.

When designing the alarm and monitoring systems with a voice evacuation system in mind, spare amplifier capacity should be included. This allows the speakers in the field to have the power settings (volume) increased without overloading the amplifier.

7.6.3 B-2 to H-6 Upgrades:

All existing audio speakers or horns not capable of broadcasting three (3) distinct evacuation signals, must be upgraded or replaced in order to conform to the H-6 code requirements.

7.7 VISUAL ALARM SIGNALING

Visual alarm signaling is performed through the use of either strobe lights, beacons, or warning signs. Strobe lights should be labeled with "ALARM," so they can be utilized with any of the three audible evacuation signals. If a fire, toxic gas or chemical spill alarm is initiated in an area, the strobe lights must be energized. The strobe lights' flashing rate should be tested to verify that it will not instigate seizures in employees with epileptic histories. As noted earlier, mounting additional strobe lights of distinctive color and labeling above each emergency spill-alarm phone in corridors containing multiple phones to aid emergency response personnel in locating the spill is recommended.

As another safety feature for employees, an additional visual alarm signaling device may be installed outside areas where entrance during an emergency condition may result in an extreme risk (flashing "DO NOT ENTER" signs). These can be connected to the alarm and monitoring system through the use of control relays or circuits. Also, in some jurisdictions, exit signs are now being required to flash in emergency conditions. If the exit signs are existing, it may be possible to mount a strobe light next to the exit sign in lieu of purchasing a replacement flashing exit sign. Local authorities should be consulted on this particular issue.

Existing visual alarm strobe lights may be utilized. However, if the lenses are silk screened with "FIRE," then they should be replaced with ones marked "ALARM." This will eliminate any confusion that might occur when an audio and visual alarm other than fire is broadcast in the area.

7.8   FIREMAN'S COMMAND STATION

Although not required by the code, a fireman's command station consisting of a telephone handset is recommended for direct communication to the ECS at remote locations of the facility. This allows communications between emergency response personnel at the scene of the emergency and the emergency control station. Pertinent information about the entire facility can be relayed to the fireman or response personnel. Facilities that use two-way radios extensively may not need this additional system if members of the plant-wide fire brigade are equipped with radios.

However, it should be noted that some of the sophisticated electronic monitoring and control equipment, as well as production tools, are very susceptible to radio frequency interference (RFI), and radio operations within a close proximity could induce false signals onto the life-safety alarm and monitoring system.

7.9   MISCELLANEOUS EQUIPMENT

Although not required by any codes, it is sometimes prudent to monitor miscellaneous pieces of process equipment for abnormal conditions. Equipment containing automatic fire extinguishing systems such as halon or CO_2 is a prime example. This must be considered in the design of the life-safety alarm and monitoring system.

7.10   HEAD END EQUIPMENT

7.10.1 Before going on to discuss the recommended type of life safety system, let us review alarm and monitoring systems.

The number of "points" in a life-safety system can become quite large. For example, consider the following:

- Fire alarm pull stations at each exit.
- Fire sprinkler flow switches for each sprinkler zone.
- Area smoke detectors.
- Recirculating air system smoke detection.
- Make-up air smoke detection, one or more per fan.
- Door holders at building separations.
- Chemical spill telephones every 150 feet in chemical delivery corridors.
- Toxic-gas monitors, low-level warning and high-level alarm per channel.
- Voice alarm speakers.
- Visual alarm strobes.
- Shutdown circuits for ventilation systems.
- Shutdown circuits for toxic gas systems.
- Monitoring of special equipment fire suppression systems (halon discharge).

Frequently, a complex custom-designed software program is necessary to "marry" the various subsystems into a cohesive life-safety system.

7.10.2   The codes do not give any requirements or restrictions as to the manufacturer or type of the life-safety system to be utilized. Therefore, the remaining portions of this chapter are guidelines provided for the reader's consideration. The reader is cautioned to carefully analyze the specific requirements of each application and select components and a system design based upon those unique circumstances. No one methodology can be expected to be a panacea.

7.10.3   Due to the size of typical semiconductor manufacturing facilities and the multiplicity and complexity of the required alarm systems, the use of a single multiplex or network proprietary central system may be appropriate.   All alarm and monitoring systems are then combined onto a central system. If all of the required alarm systems were to be wired independently, the plant would be engulfed in a mass of segregated conduits and conductors. However, through the use of multiplexing or networking, communication is conducted through only a few twisted shielded pair cables.

Multiplexing is the communication and transmission of multiple signals on a single set of conductors. Each group of devices must report its status periodically to the central processing unit (CPU). There are several manufacturers of this type of equipment: Simplex, Autocall, Edwards and Fast, to name a few.

Alarm-monitoring and control systems technology is constantly developing and changing. A vendor that has a history of developing new systems which can communicate with their own older technology should be strongly considered. Once a system is in place, modifications and additions will surely follow.

7.10.4 There is also the option of installing the system with addressable or conventional (non-addressable) devices. Addressable devices have the advantage of providing point annunciation and are frequently lower in total installed costs for H-6 facilities. Their disadvantage is, however, that a discrete point is required for each device. The Simplex 2120 Series multiplexing system contains 2040 monitoring points and 1020 signal or control points per CPU. Multiple CPUs may be interfaced utilizing Simplex's historical reporting terminal. The Edwards networked Control Panels contain provisions for 4000 points.

7.10.5 The brain of the multiplex or networked system is the CPU. It is here that all the monitoring and subsequent command decisions take place. Several CPU features should be evaluated. They include flexibility, expandability, capacity and alarm-response time.

As stated in the voice alarm evacuation section, the CPU <u>must</u> be capable of broadcasting three (3) <u>simultaneous</u> evacuation signals throughout different portions of the facility. The CPU may have to be specially configured to meet this requirement.

The CPU should also be equipped with software routines to provide control-by-event (CBE) features, whereby the receipt of an alarm or return to normal condition may be programmed to operate any or all of the control points within the system. Control-by-event programming allows emergency situations to be pre-analyzed and a response predetermined. CBE actions for life-safety functions must be retained in a non-volatile, non-erasable memory for reliability and to prevent unauthorized personnel from reprogramming vital life-safety instructions. The CPU should also be capable of retaining CBE functions programmed by the owner. These field programmed functions should be used to augment the life-safety functions. This very extensive software is custom-designed by the engineer and/or vendor.

All signaling and control circuits should have the option of being activated manually at the ECS. For the most part, however, response procedures are automatically performed by the CPU, minimizing operator involvement and the opportunity for an erroneous response.

In larger systems incorporating several separated buildings, consideration should be given to providing a stand-alone CPU for each building but still retaining reporting and remote control capability back to a central point (emergency control station). If communication/control conductors between buildings are damaged in a catastrophic event, the stand-alone CPU can control life-safety functions within its area.

7.10.6 Another recommended feature in the CPU for H-6 occupancies is the ability to program printed messages in response to an alarm condition at a given monitoring point. Not only can it be set up to provide detailed information about the alarm, but also instruct the attendant how to respond. For example, a typical alarm message might read:

HPM High Level (TLV) Toxic Gas Alarm in Module XYZ

Area:	South HPM Gas Storage Room
Gas Detected:	PH_3 (Phosphine)
Equipment:	Gas Cabinet
Location:	Exhaust Duct
Automatic Action Taken:	Gas shutdown, local alarm sounded, door holders released
Operator Action:	Call John Doe at 722-4900

To minimize data storage and future modifications, each message should be stored and recalled in multiple addressable statements. All statements and call routines must be accessible and capable of modifications with the proper authorized clearance code.

7.10.7   Emergency power must be provided for all head-end equipment, such as the CPU, as well as for all the data gathering or remote monitoring panels. If voice evacuation is utilized, the code requires battery capacity for 24 hours of system operation. It is also recommended that power be fed from a source connected to the emergency generator. Additionally, it is recommended that a UPS be provided with the emergency generator as back-up for the UPS batteries.

Additional head-end equipment at the ECS should include a bank of switches and lights for indicating and manually controlling the status of monitor and control points, CRT/keyboard for access to the CPU, printers for hard copies of detailed alarm condition response procedures, as well as graphic annunciator panels for emergency response personnel.

(Note: All information regarding the "MDA" and "Simplex Time Recorder Co." products was obtained from the latest available literature and oral communications with the factory. Each manufacturer must be contacted for current product status and operation on each new project.)

7.11 WIRING

7.11.1 All life safety alarm systems required by code should be wired in the McCullough loop (Class "A" or Style "D") configuration. Although, by code, a Class "B" system may be acceptable, the facilities' insurance carrier may require Class "A" wiring. This configuration is especially critical for the communications loop in multiplexed installations. If the forward loop is severed for any reason, (construction, explosion, fire, fault, etc.) the remote-monitoring panels can still communicate on the return loop and vice versa. Careful attention must be paid to communications loop routing, type of conductor and surge suppression.

Communication circuits between data gathering panels should be located so that they will not be in the way of local renovations in the fab.

Maximum conductor capacitance is of concern because several signaling techniques are in use that are no longer merely slow changes in DC circuits. Some, operating at effective baud rates of 4800, have bits of data that can change 4800 times per second. Maximum capacitance can effect the shape of these bits of data which can introduce errors. Maximum conductor capacitance, while normally of minimal concern, even in most high-rise buildings, presents problems of loop lengths that should be carefully reviewed in large industrial buildings.

Surge suppression must be provided:

-        for communication circuits between panels in different buildings,

- for device circuits that leave the building, and
- for control circuits to loads that develop in rush currents and counter electromotive frequency (EMF).

7.11.2 Items not required by code, such as the low-level toxic gas warning alarm and miscellaneous equipment monitoring, can be wired in accordance with Class "B" requirements.

7.11.3 Interface wiring between the life-safety alarm and monitoring systems and gas cabinets or analyzers should utilize shielded cable to protect against radio frequency and electromagnetic interferences. Generally this type of equipment is extremely sensitive and even slight interferences can cause gas cabinet shut downs and trigger false alarms. Shielded cabling prevents these interferences. At some installations, keying of hand-held radios has triggered responses in sensitive equipment.

7.11.4 All wiring must be installed in conduit.

SUMMARY

LIFE SAFETY ALARM AND MONITORING SYSTEMS (7.0)

◆   Today, numerous alarm and monitoring systems are required by codes. These include fire alarm, smoke detection, sprinkler system supervision, emergency (spill) alarm and continuous toxic gas monitoring and detection systems.

EMERGENCY CONTROL STATION (ECS) (7.2)

◆   Definition: "Emergency Control Station (ECS) is an approved location on the premises of Group H, Division 6 Occupancy where signals from emergency equipment are received and which is continually staffed by trained personnel." (UBC 406).

■   **UFC 80.301(v)   (Hazardous Materials) Storage - Supervision**
"When emergency alarm, detection or automatic fire-extinguishing systems are required in Sections 80.302 through 80.315, such systems shall be supervised by an approved central, proprietary or remote station service or shall initiate an audible and visual signal at a constantly attended on-site location."

■   When the ECS is not located in a separate building, separation from the remaining portion of the building by a two-hour wall (NFPA 72D-1.3) is required. In addition, the ECS must be constructed to conform with ANSI SE3.2 (UL 827) which includes continuous monitoring by trained and competent personnel, emergency lighting, and HVAC fed from an emergency power source or generator.

ALARM AND MONITORING SYSTEMS (7.3)

◆   Fire Alarm, Smoke Detection and Sprinkler Supervision Initiating Devices: The H-6 occupancy must be provided with either a manual or automatic supervised fire detection and alarm system installed and connected to the ECS in accordance with NFPA 72A.

■   **UFC 80.402(b)3G(vi)   (Indoor Dispensing and Use) Special Requirements for Highly Toxic and Toxic Compressed Gases - Smoke detection**
"Smoke detection shall be provided in accordance with Section 80.303(a)10."

■   **UFC 51-110(b)6 Storage of HPM within Buildings**
"A manual alarm box or approved emergency alarm-initiating device connected to a local alarm."

■   **UFC 80.301(u)   (Hazardous Materials) Storage - Emergency Alarm**
"An approved emergency alarm system shall be provided in buildings, rooms or areas used for the storage of hazardous materials."

■   **UFC 80.303(a)10   (Toxic and Highly Toxic Compressed Gases) Indoor Storage - Smoke Detection**
"An approved supervised smoke-detection system shall be provided in rooms or areas where highly toxic compressed gases are stored indoors."

◆   Types of Required Alarm Devices:

■   <u>Manual Pull Stations</u>:   NFPA Article 72A requires the distribution of manual fire alarm pull stations throughout the protected area.

- Area Smoke and Heat Detectors: Hazardous production material storage rooms are required to have a supervised smoke detection system in accordance with the above UFC requirements.

- **NFPA 70 (Article 504)**
  Classifies an "Intrinsically Safe System" as an assembly of interconnected intrinsically safe apparatus, associated apparatus, and interconnecting cables in which those parts of the system which may be used in hazardous (classified) locations are intrinsically safe circuits. An intrinsically safe system may include more than one intrinsically safe circuit.

  An "Intrinsically Safe Circuit" is a circuit in which any spark or thermal effect is incapable of causing ignition of a mixture of flammable or combustible material in air under prescribed test conditions. Test conditions are described in Standard for Safety, Intrinsically Safe Apparatus and Associated Apparatus for use in Class I, II, and III, Division 1, Hazardous (Classified) Locations, ANSI/UL 913-1988.

- **NFPA 318-7 2-3.1**
  "An air sampling or particle counter-type smoke detection system shall be provided in the cleanroom return air stream to sample for smoke."

- The installation of heat detectors is recommended in all electrical rooms.

- Duct Smoke Detectors: In all occupancies, duct-type smoke detectors are required in return air ducts for air handling units over 2,000 CFM (UMC 1009) and in the supply and return of units over 15,000 CFM (NFPA 90A). It may be unnecessary to provide supply and return detectors in the recirculating air handlers of cleanrooms (one may suffice), however, a variance of this type will require the approval of the Authority Having Jurisdiction.

- **UBC 911(b)3**

  "...Ventilation systems shall comply with the Mechanical Code except that the automatic shut-offs need not be installed on air-moving equipment. However, smoke detectors shall be installed in the circulating airstream and shall initiate a signal at the emergency control station..."

- <u>Smoke Door Holders</u>: Holders are required on doors through area separation walls (two-hour rating), building separation walls (four-hour rating) and other walls which are required to be smoke partitions.

- <u>Sprinkler Flow and Tamper Switches</u>: Sprinkler flow and tamper switches to monitor the fire sprinkler systems are required by UBC-3802 and 3803. Typically, a plant is broken into many fire sprinkler zones. The various fire sprinkler zones are each monitored separately to pinpoint the fire location.

- <u>Optical Detectors</u>: Use or dispensing of pyrophoric materials (materials which spontaneously react with air) cannot be monitored by conventional smoke or heat detectors. Their detection speed is simply too slow. Optical sensors for each application require judicious selection.

CONTINUOUS TOXIC GAS MONITORING (7.4)

♦ A continuous gas-detection system is required by Uniform Fire Code Article 51. The system must continually monitor use areas for the presence of fugitive gases that might be present.

♦ These systems must be designed in accordance with:

- UFC Article 51 and 80
- UBC Chapter 9
- Plant Safety and Insurance Requirements

◆   The following code excerpts address the requirement for gas detection systems:

■   **UFC 51.105(e)3**
"When hazardous production material gas is used or dispensed .... a continuous gas detection system shall be provided to detect the presence of a short-term hazard condition . . . The detection system shall be connected to the emergency control station."

■   **UFC 80.303(a)9   (Toxic and Highly Toxic Compressed Gases)**
"A continuous gas-detection system shall be provided to detect the presence of gas at or below the permissible exposure limit or ceiling limit.   The detection system shall initiate a local alarm and transmit a signal to a constantly attended control station (ECS).   The alarm shall be both visual and audible.
"....The gas-detection system shall be capable of monitoring the room or area in which the gas is stored at or below the permissible exposure limit or ceiling limit <u>and the discharge from the treatment system</u> at or below one-half the IDLH limit."

◆   **Automatic shutdown** of the gas cabinet supply lines is required. The following code excerpts address the requirement for automatic shut-off valves.

■   **UFC 51.107(b)2**
■   **UFC 80.402(b)3F(v)**
■   **UFC 80.402(c)8D**

◆   Sensor Locations: The codes do not specify the exact location of gas sensors inside the fab area; they simply state that the gas must be detected.   Extremely high air movement rates play a major role on sensor location and spacing.

♦   Control, Signaling and Levels of Detection: We recommend that two levels of detection be provided, a low warning level set at 1/2 TLV (threshold limit value) and a high alarm level set equal to TLV (code-required).   If the high level is attained, an alarm is sent to the ECS, a distinct toxic gas evacuation signal must be broadcast locally throughout the area, and the detected gas supply line must be automatically shut-off.

♦   Gas Analyzers: The definition of "continuous monitoring" requires that each sensor or location be sampled and evaluated in periods not exceeding 30 minutes.   Life-safety systems are required to be electrically supervised to provide indications of opens and shorts in wiring or malfunctioning components.

   ▪   All gas analyzers should be located outside the fab.

   ▪   It is recommended that high traffic areas be monitored more frequently than unmanned storage areas.

♦   Power Requirements: Connection to an emergency power source is requisite for all HPM gas analyzers. If the particular analyzer utilized must go through a "boot-up" process each time it is powered up, a UPS (uninterruptible power system) should be provided.

♦   B-2 to H-6 Upgrade Phasing: Toxic gas monitoring systems should be immediately upgraded to conform to all the new applicable codes. Not only does it protect employees but it also reduces the facility's liability.

EMERGENCY SPILL ALARM (7.5)

♦   In new H-6 facilities, transporting HPMs in code-required exit corridors is prohibited. In existing facilities, the Uniform Fire Code will allow transportation of HPMs in the exit corridors only if spill-alarm signaling devices are provided at 150-foot intervals (among other requirements outlined in other sections).

- The purpose of the HPM emergency spill alarm is to initiate a local alarm and to report to the ECS, the conditions of the spill, leak or other accident within the exit corridor used to transport HPMs.

- Although a manual pull station is acceptable by the code, it is recommended that utilizing a fireman's telephone mounted in an enclosure be considered.

◆ Spill Containment Areas:

- **UFC 80.301(l)4**
  ". . . When secondary containment is required, a monitoring method capable of detecting hazardous material leakage from the primary containment into the secondary containment shall be provided."

  - Another good practice for detecting chemical spills is to install detection systems in the bottom of piping trenches below the fab area's raised floor.

AUDIBLE ALARM EVACUATION (7.6)

◆ In lieu of the "distinct tones" acceptable to the codes, it is recommended that you provide a pre-programmed voice alarm system. The voice alarm evacuation system consists of high-reliability speakers located throughout the facility which will broadcast distinct verbal messages to reduce confusion and help eliminate panic in an emergency evacuation of a given area. The following distinct audible alarm systems are required:

- Fire alarm evacuation signal.
- Toxic gas evacuation signal.
- Chemical spill evacuation signal.

♦   If a voice evacuation system is not provided, all existing audio speakers or horns not capable of broadcasting three (3) distinct evacuation signals, must be upgraded or replaced in order to conform to the H-6 code requirements.

VISUAL ALARM SIGNALING (7.7)

♦   If a fire, toxic gas or chemical spill alarm is initiated in an area, strobe lights must be energized. The strobe lights must be of a frequency which will not induce epileptic seizures.

♦   Strobe lights of distinct color should be mounted above emergency spill alarm phones.

♦   Flashing "DO NOT ENTER" signs should be considered for all hazardous areas. Connect to the alarm/monitoring system.

♦   Flashing exit signs may be required. Check with your local jurisdiction.

FIREMAN'S COMMAND STATION (7.8)

♦   Although not required by the code, a fireman's command station consisting of a telephone handset is recommended for direct communication to the ECS at remote locations of the facility.

HEAD END EQUIPMENT (7.10)

♦   Due to the size of typical semiconductor manufacturing facilities and the multiplicity and complexity of the required alarm systems, the use of a single multiplex or network proprietary central system may be appropriate. All alarm and monitoring systems may then be combined within such a system.

♦   Multiplexing is the transmission and communication of multiple signals on a single set of conductors.   Each device reports its status periodically to the CPU.

♦   Addressable devices, which are discrete points for each device may be desirable, in lieu of zoned devices; however, the number of required points and inherent complexity of the system must be evaluated.

♦   A stand-alone CPU may be desirable for each building or major zone of a facility to limit the potential liability of a single CPU.   The stand-alone CPUs must report to the supervisory or "host" CPU at the ECS for monitoring and external commands.

♦   Emergency power must be provided for all head-end equipment, such as the CPU, as well as for all the data gathering or remote monitoring panels.   It is recommended that a UPS be provided with the emergency generator as back-up for the UPS batteries.

WIRING (7.11)

♦   All life safety alarm systems required by code should be wired in the McCullough loop (Class "A" or Style "D") configuration.   If the forward loop is severed for any reason, the remote-monitoring panels can still communicate on the return loop and vice versa.

♦   Conductor capacitance is of concern because of signaling techniques used in the alarm systems.   Conductor capacitance may present problems where loop lengths are extreme, therefore, the design of the wiring system must be carefully reviewed in large industrial buildings.

♦   Interface wiring between the life-safety alarm and monitoring systems and gas cabinets and other devices or analyzers should utilize shielded cable to protect against radio frequency and electromagnetic interferences.

♦   All wiring must be installed in conduit to protect it from external damage.

8

Retrofit and Renovation of Wafer Fabs to Comply with Hazardous Occupancy Codes

8.1 INTRODUCTION

This chapter is intended to outline a method of developing retrofit/renovation projects which will result in compliance with the intent of the UBC/UFC "code family" for hazardous occupancies. In that the implications of the code requirements represent significant expenditures with the potential for interruption of production, it may be desirable to accomplish the work in a phased manner.

Despite the fact that the UBC may allow an existing B-2 industrial facility to continue operating without upgrade, it is incumbent on the owner to ensure the facility is safe. One of several code references in this regard is cited:

> **UBC 104(c) Application to Existing Buildings and Structures - Existing Installations**
> "Buildings in existence at the time of the adoption of this code may have their existing use or occupancy continued, if such use or occupancy was legal at the time of the adoption of this code, provided such continued use is not dangerous to life . . ."

Thus, whether or not the local authorities have exerted pressure on an owner, it is necessary to continually evaluate the life-safety of each facility and act accordingly. It is obvious that a catastrophic (or lesser) incident is to be avoided at all costs. It is equally obvious that recent code provisions, changes in technology and changes in the use of a facility may result in the need to reconsider whether or not the facility is "dangerous to life." Facilities which were previously considered safe, may no longer be so considered.

It is not the intent of this book to offer definitive guidelines as to when an owner should or must implement code upgrades. These are evaluations best undertaken by the owner's risk management experts, in concert with the insurance underwriter and the local code authorities. However, once the decision has been made to upgrade the facility, the following discussion should prove helpful.

8.2   CODE AGENCY LIAISON

8.2.1   Introduction:

A good working relationship with the local code authorities is desirable, if not mandatory, to operation of a hazardous occupancy facility. Building officials have significant concern over the safety of the public (and their own personal liability) related to the operation of a facility which utilizes toxic, flammable, corrosive, explosive or otherwise hazardous materials.

Recent life-threatening or life-taking incidents involving hazardous materials (e.g., Union Carbide at Bhopal, India; Three Mile Island; and others) have served to heighten the awareness and concern of building officials.

8.2.2   Plan Check and Permit Process:

The process of identifying existing system or facility deficiencies, evaluating the means to correct them, obtaining the necessary permits prior to implementation and then constructing the retrofits is a complex and important one. There are significant costs associated with compliance with the codes. A carefully-orchestrated program is necessary to insure successful upgrade of the facility. Because the needs in each facility are varied, it is not feasible to offer a prescriptive approach to compliance. The following discussion is intended to assist the user in developing a plan for each unique project, based on sound rationale.

Many industrial facilities operate under a "Registered Plant Program." In essence, the Registered Plant Program (RPP) allows the owner of an industrial facility to perform certain maintenance, modifications and upgrades of the facility without obtaining a specific permit for the work. The purpose of this program is to avoid unduly hampering the progress of minor renovation and remodel, with the delay inherent in obtaining a permit. The provisions of the typical Registered Plant Program require the plant operator to have on staff (or retain as a consultant) a professional engineer to serve as the responsible party for ensuring that renovation work is performed in a manner compatible with codes and good practice. The registered plant provisions generally have restrictions on their application. Permits are generally required for work related to fire separations, building expansions, structural modifications or other activities which affect life-safety. The RPP generally has provisions for annual or periodic reporting of the activities which have transpired in order to update the building official on the status of the facility.

While it may be tempting to circumvent the permitting process for certain code compliance upgrades, under the guise of the RPP, it is important not to overstep the bounds of the "plant engineer's" jurisdiction. We believe it is in the owner's best interest to maintain a relationship of trust and rapport with the building official. The type of projects which fall into the category of requiring a hazardous occupancy upgrade need to be reviewed with the building official and subjected to the permitting process if deemed appropriate by the building official.

The following is an outline of a reasonable means of dealing with the problems a facility owner faces with the advent of ever-tightening code requirements.

Building officials will generally understand the problems faced by owners trying to comply with the new codes, if those problems are clearly and concisely stated by the owner. As a result, they are usually willing to cooperate in the implementation of reasonable measures to comply with code intent, <u>if they believe the owner is acting in good faith</u>.

The UBC provides the building official latitude in dealing with difficult or unusual circumstances under Sections 105 (Alternate Materials and Methods of Construction) and Section 106 (Modifications). The UFC has similar provisions under Sections 2.301(a) and 2.301(b). (*Refer to Chapter 1 of this book for further discussion.*)

These code sections give the owner/designer the opportunity to negotiate compliance strategies which may not meet the letter of the code, but satisfy its intent.

In the unlikely event that a satisfactory solution to a specific problem cannot be resolved with the building official, the UBC provides a procedure via Section 204 which establishes a Board of Appeals consisting of independent, knowledgeable citizens who may render decisions concerning the suitability of alternate materials and methods. Generally, a Board of Appeals may approve alternate compliance strategies; however, they may not circumvent the intent of the code with respect to life-safety. The UFC makes similar provision for a Board of Appeals in Section 2.303.

8.2.3   Preliminary Code Evaluation/Study:

It is imperative that the owner conduct a thorough evaluation of existing facilities which are known to deviate from the requirements of the UBC/UFC Code family. This engineering/architectural evaluation must address all of the pertinent life-safety issues raised by the codes <u>to establish a baseline upon which to build a program for compliance.</u>

The basic areas of concern include:

- Exiting. *BLDG. DEPT.*
- Area and Occupancy Separations. *BLDG. DEPT*
- HPM Storage and Handling. *FIRE DEPT.; CLASSIFIES BLDG.*
- Ventilation.
- Monitoring and Alarm Systems.
- Emergency Operation Scenario.
- Fire Suppression Systems.

(The Author has found it helpful to utilize a code evaluation checklist to ensure the study thoroughly addresses all of the issues. In this regard, he has developed a 20-page survey format to organize the important criteria. For further information or assistance, you may contact the Author.)

The code evaluation should culminate in a report to management which outlines the deficiencies of the facility, addresses potential compliance strategies, quantifies potential cost and recommends a plan of action. The report should address the issue of the required or legally-dictated level of compliance which must be achieved, versus the level of compliance which should be accomplished to meet corporate guidelines and safety objectives. In other words, some code requirements may not be mandatory in a retrofit application, as a result of a "grandfather" clause, but should be done, nevertheless, to limit corporate liability and mitigate danger to life and property.

The code evaluation/study will prepare the owner for the preliminary presentation to the building official discussed below. Disclosure of all of the information contained in the code evaluation may be a concern to the owner due to its sensitive nature; however, such disclosure is required by law (UFC 2.302) where it is a matter of personnel safety.

8.2.4   Preliminary Presentation of Project to Building Officials:

Early on in the design or review process, a dialogue should be established between the owner/designer team and the building official. The designer must have accomplished a preliminary evaluation of the needs of the owner for the retrofit project and the implications of the codes concerning those needs. The owner/designer needs to approach the building official with a clear plan, outlining the scope of the project and his understanding of the code requirements relevant to that scope.

Be prepared with intelligent, considered questions related to issues which are subject to interpretation and the use of alternative methods or materials. The building official will generally render opinions (at least in concept) concerning these issues. If the issues are complex, or new to the jurisdiction (as many hazardous occupancy issues are), the building official or fire department representative may request a period of time to review the situation and confer with staff, outside experts, the International Conference of Building Officials (ICBO), or other technical advisers. UFC Section 2.302 states that "the chief" may require the owner of a facility to provide a technical opinion or report on a particularly complex issue, at no expense to the officials.

The approach to this type of meeting should be to "put all of the cards on the table" with a professionally-prepared, written narrative. This narrative outlines the goals of the project and the issues involved in achieving the goals of the owner, as well as code compliance. The narrative should be supplemented by drawings, sketches, diagrams and/or calculations, when appropriate to the complexity and level of progress on the project. The more (professionally developed) information the building official has to evaluate, the greater his level of confidence will be in the project. With a high level of confidence, the building official has less tendency to render a "conservative" opinion or ruling in matters of judgment and interpretation. This open and honest approach is generally most advantageous to the owner.

Issues which the owner/designer should be prepared to discuss depend on the scope and complexity of the project, however, some examples are offered:

- General size and arrangement of the facility and its relationship to the site and other buildings.
- Provisions for access to all sides of the building.
- Separation of the building from adjacent buildings.
- Size, arrangement and separation of various occupancies within the facility.
- Location and arrangement of provisions for exiting.
- Location and size of the hazardous occupancies within the overall project.
- Location and size of HPM storage areas with respect to other occupancies, property lines and roads.
- Proposed means of delivery of hazardous materials to the facility.
- Arrangement and usage of multiple levels of the fab.
- Location of the emergency control station.
- Proposed method of handling emergency power and ventilation.
- Proposed method of dealing with alarm and monitoring systems.
- Proposed method of "treatment" of toxic or hazardous discharges and their location with respect to property line and fresh-air intakes.

The narrative should indicate the plans of the owner to phase compliance strategies (if applicable) in a logical and timely manner, based on the rationale that full and immediate compliance with the new code provisions will be prohibitively expensive or potentially cause a shut down of operations. The rationale of "moving in the direction of full code compliance" should be looked upon favorably by the building official. It may be noteworthy to mention that the facility would not need to be upgraded to current code requirements, if a revision were not contemplated (reference UBC Section 104(c)).

We have found it advantageous to offer implementation of strategies which may not be strictly required to comply with the code, in lieu of code requirements which are difficult or impossible to achieve. Where such non-mandatory strategies can be shown to improve life safety and qualify under the terms of Section 105, Alternatives, or Section 106, Modifications, the building official should consider approval of such substitute methods.

The owner/designer should plan to leave the written and graphic materials with the building official, after the in-person presentation, for his study and review. Pressuring the building official for rulings on issues he is uncomfortable with may back-fire and tend to strain the relationship of trust necessary to acceptance of a project of this type.

The owner/designer should follow-up the meeting with the building official with a formal memorandum outlining the issues discussed, issues resolved, concepts agreed upon and issues pending further resolution. Where appropriate, reference specific code provisions to ensure thorough understanding. It may be appropriate to document your rationale on a position which the building official has ruled against and ask for further review. Document all decisions which were made concerning alternative methods/materials or modifications to show clearly the rationale for their acceptance. The letter should request an early <u>written</u> reply to confirm the designer's documentation.

The benefits of early understanding of the project requirements are:

- Faster plan check.
- Less coordination difficulty.
- Greater assurance of obtaining permit.
- Improved rapport with building official.
- Less design time.
- Earlier start of construction.

8.3   PRIORITIZING CODE COMPLIANCE ISSUES

8.3.1   General:

It should be obvious by now that the intent of the Uniform Building Code, the Uniform Fire Code and companion regulations is to enhance the safety of facilities using hazardous materials. It is equally obvious that converting an existing B-2 Occupancy fab to fully-conforming "H" Occupancy might result in curtailment, if not virtual shutdown, of production or operations. Furthermore, the costs associated with full compliance with the codes may be prohibitive, and, thus, not feasible for an owner to assume at one time. Therefore, it may be necessary to prioritize the compliance issues and proceed in following a phased retrofit plan which delivers the greatest life-safety benefit for the money spent.

Compliance issues should be prioritized on the basis of the perceived danger to life and property posed by each issue, if noncompliance is continued. One method of evaluating the potential danger of each issue and, therefore, its priority in the phased plan, is to prepare a "risk matrix" as proposed below.

8.3.2   Compliance Evaluation Matrix:

Figure 8.1 is a sample format of an evaluation matrix. The issues shown are relatively consistent for typical wafer fabs; however, the reader is cautioned to recognize that each project is unique; therefore, these are offered as a starting point in the evaluation process. Each project may have many more critical issues to consider. The relative level of compliance of the existing facility to the new codes on each issue will vary for each retrofit project.

The method of "weighing" the relative importance of each criteria with respect to each of the code compliance issues is highly subjective. The weighing of each issue should be done with input

PRIORITIZATION DECISION MATRIX (NOTE: ASSIGN WEIGHT (1-5) TO EACH CONSIDER-ATION AND GRADE (1-5) TO EACH ISSUE. MULTIPLY WEIGHT x GRADE FOR SCORE IN EACH, THEN SUM)	CONSIDERATIONS AND CONCERNS													
	RELATIVE IMPORTANCE TO LIFE SAFETY	CURRENT LEVEL OF DEFICIENCY	CODE AGENCY "RED TAG"	TIME TO IMPLEMENT	RELATION TO OTHER ISSUES	EASE OF IMPLEMENTATION	COST OF IMPLEMENTATION	"SPIN- OFF" BENEFITS	IMPACT ON PRODUCTION				TOTAL SCORE	PRIORITY RANK
COMPLIANCE ISSUES WEIGHT														
EMERG. CONTROL STATION														
TOXIC GAS MONITORING														
SPILL ALARM/MONITORING														
FIRE DETECTION/ALARM														
EXIT DOORS														
EXIT CORRIDORS														
OCCUPANCY SEPARATION														
EXHAUST RATE COMPLIANCE														
EMERGENCY POWER														
CHEMICAL STORAGE ROOMS														
TOXIC GAS STORAGE/DISPENSE ROOMS														
FIRE SPRINKLER SYSTEM COMPLIANCE														
EMERGENCY SHUT-OFFS (HPM'S + AHU'S)														
TREATMENT OF TOXIC RELEASE (GASES)														
CONTAINMENT OF CHEMICAL SPILLS														
SEGREGATE AIR HANDLING SYSTEMS														
SEPARATE ACID/FLAMMABLE EXHAUST														
FIRE DAMPERS AND DRAFT STOPS														

Figure 8.1 PRIORITY DECISION MATRIX may be used as a tool to subjectively and somewhat quantitatively prioritize needed code compliance retrofits.

from safety/risk management staff, insurance representatives and code officials. While an absolute consensus may not be feasible, we believe that a reasonable level of agreement can be obtained.

8.3.3 Our experience with wafer fab retrofit projects is that the following priority list may be reasonable:

1.   Provide proper exiting from all occupied areas.

2.   Provide a comprehensive fire-alarm and detection system.

3.   Provide toxic gas monitoring and alarm systems with appropriate shut-off of HPM delivery.

4.   Provide emergency power for exhaust, make-up air ventilation and other life-safety systems.

5.   Provide an emergency control station.

6.   Construct area and occupancy separation walls to limit fab areas to required maximum. (This is generally relatively inexpensive).

7.   Provide fire-rated doors and windows in fire-rated partitions.

8.   Provide fire sprinklers throughout all areas of the fab and exhaust-duct systems.

9.   Remove chemical storage cabinets in exit corridors and replace, where required, with fire-rated pass-throughs with necessary ventilation.

10.   Isolate HPM storage areas from the fab proper in suitable H-1, H-2, H-3 or H-7 storage rooms with appropriate occupancy separation walls and independent ventilation systems.

11.   Provide fab ventilation and power-off systems.

12.   Provide leak/spill detection and alarm (telephone) systems.

13.   Provide fire dampers in supply ducts which penetrate fire-rated partitions.

14.   Provide draft stops to isolate multiple levels of a common fab.

15.   Isolate the air handling systems for corridors from the fab air handling systems.

16.   Provide proper exhaust ventilation for all levels of multiple-level fabs. (This may be construed by code agencies to mean that one cfm per square foot of exhaust and make-up air is required at each level of the multiple level fab). This equipment must be on emergency power.

(Note: The exhaust rates in most existing wafer fabs generally comply with the minimum requirements of current codes as a result of process exhaust requirements. The designer/evaluator needs to ensure that each area within the fab, however, meets the minimum ventilation rate requirement.)

17.   Segregate acid and flammable exhaust-duct systems.

18.   Provide "treatment" systems for toxic gas storage facilities.

19.   Others as may develop for each project. . . .

A WORD OF CAUTION: The above listing and order is not all inclusive or appropriate for each project! It is to be used as an example only. Each retrofit project must be evaluated on its own merit and a prioritized list for each project specifically developed by the owner/designer/official team!

8.4   PROJECT PHASING

8.4.1   General:

Construction phasing of code upgrades is crucial to several issues which will influence the success of the project:

■   Life safety.
■   Cost of retrofit.
■   Impact on production.

It is important to have a clear understanding of the overall project process during the conceptual design phase so that all parties to the project understand its impact on their jurisdiction and responsibilities. The relevant groups include, but may not be limited to:

- User/Production Group.
- Safety Department.
- Facilities Engineering Group.
- Environmental Compliance Group.
- Code Authorities.
  - Building Department
  - Fire Department
  - Air and Water Quality Regulators
- Purchasing/Financial Group.
- Insurance Underwriter.
- Security Department.

Each group should be fully informed of the project scope, so they may "buy into" the proposed process at the conceptual design stage. Careful planning of phasing and anticipated schedules will help all parties understand such issues as:

- Production interruption.
- Potential hazards.
- Cash flow.
- Impact on facility infrastructure.

8.4.2 Project Manager:

It is essential for the owner to assign a project manager to each project to help ensure its success. The project manager functions as the "clearing house" for all information and decisions. The project manager is the one point-of-contact for all parties interested in the project. The owner's project manager may also be the design project manager if the construction documents are prepared in-house. If an engineering/architecture firm or group is retained to prepare the design and construction documents, then the owner's project manager is the point-of-contact with the design team project manager. The project manager serves to focus all members of the team on goals, schedules, budget and related concerns.

One point-of-contact between owner, designer, contractor and building official should be maintained throughout the course of the project, until completion, to ensure important issues do not "fall into a crack." The project manager should be responsible for maintaining the record copy of all documents, meeting minutes, directives, estimates, rulings, etc. for the protection of the owner. Such records will be crucial in the event of an instance of "convenient" memory loss on anyone's part.

The project manager's team needs to include all or some of the following resources:

- Engineer/Architect (may be owner's staff).
- User Group Representative/Spokesman.
- Facility Engineering Group.
- Operations Group.
- Safety Representative.
- Environmental Compliance Representative.
- Security Representative.
- Construction Manager (may be owner's staff).
- General Contractor (if appropriate).
- Purchasing Agent.
- Construction Coordinator (in-house staff representative).

8.4.3   Scheduling:

Scheduling of all phases of the project is crucial to the successful outcome. Whether this task is accomplished by an outside engineering/architectural (E/A) firm, in-house facility engineering staff or a construction manager, preparation of the schedule must address the needs of the production group to coordinate necessary interruptions in their operation. The building official will also need to be advised of the plan, particularly if multiple bid packages and multiple permits are desired.

Production shut-downs must be carefully coordinated with the user group to minimize impact on revenue! The need for shut-downs must be a joint decision made by the owner, designer, user and contractor/construction manager to ensure all goals of the project are met and construction is feasible.

Scheduling must take into account adequate time for the following tasks:

- Conceptual Design.
- Code Analysis/Evaluation and Development of Strategy.
- Estimating.
- Review by All Interested Groups.
- Code Agency Review.
- Design Development and Preparation of Construction Documents.
- Funding.
- Equipment and Material Purchase and Delivery.
- Bidding and Award.
- Permitting.
- Allowable "Windows" of Production Shut-down.
- Construction Time.
- Equipment and System Start-Up and Commissioning.
- Move-in and Acceptance.
- Production Equipment Hook-up.
- Final Acceptance and Certificate of Occupancy.

Obviously, the time to complete a project from conception to approval and beneficial use can be significant. Depending on the scope, a period of six to 18 months or more can be expected. In order for the project to flow smoothly, the critical path for decisions and construction must be evaluated periodically and faithfully followed.

We have found that in projects where time is of the essence (and what project does not fall into this category?), it is often advantageous to pre-purchase the major components or systems before the construction documents are completed. Some of the equipment and systems which should be considered for pre-purchase include:

- Emergency generator and fuel tank.
- Electrical switchgear and transformers.
- Air handling and exhaust equipment.
- Alarm/monitoring components.
- Automation/control components.
- Treatment/scrubbing components.

8.4.4   Production of Construction Documents:

Phased construction frequently means a complex set of bid and permit packages.   The process may best be administered by an outside construction manager or a knowledgeable member of the in-house staff (provided this staff member can dedicate his time to the project). The construction manager or individual filling this role should be on-board as early in the process as feasible, to help with value engineering, planning, estimating and liaison of all parties.

A typical breakdown of bid packages for a project may include:

- Demolition.
- Pre-purchase of Equipment and Materials.
- "Tie-in Package" (connection to existing utilities and services).
- Architectural Revisions.
- Infrastructure Upgrades.
- Mechanical Air Handling.
- Process Piping.
- Fire-suppression System.
- Life-safety Alarm/Monitoring.
- Electrical Power (including emergency power).
- Controls and Automation.

General Conditions:

While there is no substitute for a contractor who has worked in the plant before and performed in a satisfactory manner, the contract must clearly state the obligations and requirements. We have found that the best place to clearly set forth the project protocol and administrative requirements is in the "boiler plate" section of the specifications known as the "Supplementary General Conditions" (SGC). The SGC should correspond to a standard document such as the General Conditions of the Contract published by the American Institute of Architects (AIA) or the Engineers Joint Construction Document Council (EJCDC) version of the General Conditions. The SGC must be carefully prepared by the owner or the consultant, and the contractor must be made aware of the importance of the document during the pre-bid conference.

The "tie-in" package(s) allow(s) future connection to utilities and infrastructure with minimal disruption of the production process. The tie-in package should anticipate, as much as possible, the need to connect to utilities (steam, water, compressed air, power, etc.) air-handling and exhaust systems, openings in construction for future access and provision of temporary facilities to accommodate later construction needs.   The tie-in package is usually accomplished during a planned facility shutdown, such as during the Christmas holiday, Easter holiday, Memorial Day, etc. Final connection to the utility or system is usually made during a weekend and short-duration (four hours or less) shut-down.

Infrastructure upgrades may include such utility system revisions as installation of an emergency generator, upgrade or installation of the emergency control station, fire alarm for the plant, modifications to the electrical service to provide redundant feeders, enhancement of the chilled water supply, structural support for equipment, etc. Infrastructure upgrades can be constructed before, or at the same time as, the primary project. The key criteria is that the utility systems be available to handle the loads of the revised facility systems when they are due to come "on-line." The infrastructure upgrades may take longer to implement than the actual work in the facility proper; therefore, critical path planning is essential to success of the project.

8.4.5   Construction:

Construction within the facility, or on systems which support the facility and its processes, must be carefully documented with detailed drawings and specifications by the designer. It is vitally important that the contractor(s) selected to perform the work understand the critical nature of the operation of a manufacturing operation and the impact their activities may have on production. The contractor(s) must be advised of his obligations to cooperate with the owner and pay particular attention to the following areas of concern:

- Vibration caused by construction activity.
- Generation of dust and dirt.
- Noise generated by construction.
- Interruption of utilities (not to be tolerated).
- Maintenance of environmental conditions.
- Cleanroom protocol and procedures (if he must enter the clean space to work).
- Safety related to hazardous materials.
- Allowable "shut-down" windows and the potential for these dates and times to be revised.
- Overtime and premium time work to expedite completion.

The construction manager or owner's construction coordinator should insist on periodic progress meetings with the contractor(s) to update the schedule, discuss potential delays or problems and generally insure a successful project through coordination and cooperation. The requirements for any required utility outages should be discussed and a detailed plan prepared to ensure that all contractors' and owner's support groups are aware of their responsibilities and the potential implications of the shutdown.

8.4.6   System Start-Up and Commissioning:

System start-up and commissioning must be carefully accomplished under the direction of the designer. This phase includes:

- Testing, adjusting and balancing of air-handling and exhaust systems.
- Proof of all control sequences.
- Load test of emergency generator.
- Proof of all alarm and monitoring systems.
- Testing and acceptance by code authorities.
- Receipt of certificate of occupancy.

The importance of the start-up and commissioning phase should be obvious, yet on many projects this work is left to the contractor to oversee and verify in an unstructured and sometimes casual manner. Additionally, the time required to bring complex systems on-line is often underestimated, thus resulting in schedule delays. We feel very strongly that both the owner and designer must be actively involved in the activity to ensure the success and life safety of the project.

Periodic inspection and testing of facility life-safety systems is mandated by codes in some instances. The owner must be fully aware of the various requirements of the UFC, State Fire Marshal, local building official,   insurance carrier,   etc.   In addition,   safety department and risk management experts should evaluate each facility and establish schedules for comprehensive, periodic testing and adjustment.

Retrofit of an existing facility to conform to current code requirements can be an extremely challenging endeavor. The success of the project is dictated by the level of planning, understanding and commitment of all concerned.   A successful retrofit can be accomplished and will benefit the owner by providing a safer environment in which to work. We hope this book is an aid in your success, and wish you the best!

SUMMARY

RETROFIT AND RENOVATION OF HAZARDOUS FACILITIES TO COMPLY WITH "H" CODES (8.0)

♦   In that the implications of the code requirements represent significant expenditures with the potential for interruption of production, it may be desirable to accomplish the work in a phased manner.

♦   It is incumbent on the Owner to ensure the facility is safe. Thus, it is necessary to continually evaluate the life-safety of each facility and act accordingly.

■   **UBC 104(c)**
   "Buildings may have their existing use or occupancy continued, provided such continued use is not dangerous to life . . ."

CODE AGENCY LIAISON (8.2)

♦   A good working relationship with the local code authorities is desirable, if not mandatory, to operation of a hazardous occupancy facility.

♦   Plan Check and Permit Process: The UBC provides the building official latitude in dealing with difficult or unusual circumstances under Sections 105 (Alternate Materials and Methods of Construction) and Section 106 (Modifications).   These code sections give the owner/designer the opportunity to negotiate compliance strategies which may not meet the letter of the code, but satisfy its intent.

♦   Preliminary Code Evaluation/Study: Conduct a thorough evaluation of existing hazardous facilities which are known to deviate from the requirements of the UBC/UFC Code family.

◆   The code evaluation should culminate in a report to management which outlines the deficiencies of the facility, addresses potential compliance strategies, quantifies potential cost and <u>recommends a plan of action.</u> Some code requirements may not be mandatory, but should be done, nevertheless, <u>to limit corporate liability and mitigate danger to life and property.</u>

◆   Early on in the design or review process, a dialogue should be established between the owner/designer team and the building official.

■   The Owner should discuss plans to phase compliance strategies (if applicable) in a logical and timely manner, based on the rationale that full and immediate compliance with the new code provisions will be prohibitively expensive or potentially cause a shut down of operations.

■   The owner/designer should follow-up to the meeting with the building official with a formal memorandum outlining the issues discussed, issues resolved, concepts agreed upon and issues pending further resolution.

PRIORITIZING CODE COMPLIANCE ISSUES (8.3)

◆   Compliance issues should be prioritized on the basis of the perceived danger to life and property posed by each issue, if noncompliance is continued. A decision-making matrix (**Figure 8.1**) may help you quantify or weigh the relative importance of each issue to your facility.

PROJECT PHASING (8.4)

◆   The project manager serves to focus all members of the team on goals, schedules, budget and related concerns.

♦   Scheduling: Preparation of the schedule must address the needs of the production group to coordinate necessary interruptions in their operation. The building official will also need to be advised of the plan, particularly if multiple bid packages and multiple permits are desired.

- It is often advantageous (or necessary) to pre-purchase the major components or systems before the construction documents are completed.

- It is vitally important that the contractor(s) selected to perform the work understand the critical nature of the operation of the industrial facility and the impact their activities may have on production.

- System start-up and commissioning must be carefully accomplished under the direction of the designer.

- The time required to bring complex systems on-line is often underestimated, thus resulting in schedule delays.

- The success of the project is dictated by the level of planning, understanding and commitment of all concerned!

9

Uniform Building Code Definitions and Abbreviations

The following pages are directly from Chapter 4 (Definitions and Abbreviations) of the 1991 Uniform Building Code.

Definitions

Sec. 401. (a) **General.** For the purpose of this code, certain terms, phrases, words and their derivatives shall be construed as specified in this chapter. Words used in the singular include the plural and the plural the singular. Words used in the masculine gender include the feminine and the feminine the masculine.

Where terms are not defined, they shall have their ordinary accepted meanings within the context with which they are used. *Webster's Third New International Dictionary of the English Language, Unabridged,* copyright 1986, shall be considered as providing ordinarily accepted meanings.

(b) **Standards of Quality.** The standards listed below labeled a "U.B.C. standard" are also listed in Chapter 60, Part II, and are part of this code. The other standards listed below are guideline standards and as such are not adopted as part of this code (see Sections 6002 and 6003).

1. **Noncombustible material**

A. U.B.C. Standard No. 4-1, Noncombustible Material Test

2. **Burning characteristics of building materials**

A. U.B.C. Standard No. 42-1, Test Method for Surface-burning Characteristics of Building Materials

B. U.B.C. Standard No. 25-28, Fire-retardant-treated Wood Tests on Durability and Hygroscopic Properties

3. **Corrosives and irritants**

A. Code of Federal Regulations 49, Part 173, Appendix A, Testing for Corrosiveness

B. Code of Federal Regulations 16, Sections 1500.41 and 1500.42, Methods of Testing Primary Irritant Substances and Test for Eye Irritants

4. **Ranking of hazardous materials**

A. U.F.C. Standard No. 79-3, Identification of the Health, Flammability and Reactivity of Hazardous Materials

5. **Classification of plastics**

A. ASTM D 635, Method of Test for Determining Classification of Approved Light-transmitting Plastics

A

Sec. 402. ACCESS FLOOR SYSTEM is an assembly consisting of panels mounted on pedestals to provide an under-floor space for the installations of mechanical, electrical, communication or similar systems or to serve as an air-supply or return-air plenum.

ACI is the American Concrete Institute, Box 19150, Redford Station, Detroit, Michigan 48219.

ADDITION is an extension or increase in floor area or height of a building or structure.

AEROSOL is a product which is dispensed by a propellant from a metal can up to a maximum size of 33.8 fluid ounces or a glass or plastic bottle up to a size of 4 fluid ounces, other than a rim-vented container.

AGRICULTURAL BUILDING is a structure designed and constructed to house farm implements, hay, grain, poultry, livestock or other horticultural products. This structure shall not be a place of human habitation or a place of employment where agricultural products are processed, treated or packaged; nor shall it be a place used by the public.

AISC is the American Institute of Steel Construction, Inc., 400 North Michigan Avenue, Chicago, Illinois 60611.

ALLEY is any public way or thoroughfare less than 16 feet but not less than 10 feet in width which has been dedicated or deeded to the public for public use.

ALTER or **ALTERATION** is any change, addition or modification in construction or occupancy.

AMUSEMENT BUILDING is a building or portion thereof, temporary or permanent, used for entertainment or educational purposes and which contains a system which transports passengers or provides a walkway through a course so arranged that the required exits are not apparent due to theatrical distractions, are disguised or not readily available due to the method of transportation through the building or structure.

ANSI is the American National Standards Institute, 1430 Broadway, New York, New York 10018.

APARTMENT HOUSE is any building or portion thereof which contains three or more dwelling units and, for the purpose of this code, includes residential condominiums.

APPROVED, as to materials and types of construction, refers to approval by the building official as the result of investigation and tests conducted by the building official, or by reason of accepted principles or tests by recognized authorities, technical or scientific organizations.

APPROVED AGENCY is an established and recognized agency regularly engaged in conducting tests or furnishing inspection services, when such agency has been approved.

APPROVED FABRICATOR is an established and qualified person, firm or corporation approved by the building official pursuant to Section 306 (f) of this code.

AREA. See "floor area."

ASSEMBLY BUILDING is a building or portion of a building used for the gathering together of 50 or more persons for such purposes as deliberation, education, instruction, worship, entertainment, amusement, drinking or dining or awaiting transportation.

ASTM is the American Society for Testing and Materials, 1916 Race Street, Philadelphia, Pennsylvania 19103.

ATRIUM is an opening through two or more floor levels other than enclosed stairways, elevators, hoistways, escalators, plumbing, electrical, air-conditioning or other equipment, which is closed at the top and not defined as a mall. Floor levels, as used in this definition, do not include balconies within assembly occupancies or mezzanines which comply with Section 1717.

AUTOMATIC, as applied to fire-protection devices, is a device or system providing an emergency function without the necessity of human intervention and activated as a result of a predetermined temperature rise, rate of rise of temperature or increase in the level of combustion products.

B

Sec. 403. BALCONY is that portion of the seating space of an assembly room, the lowest part of which is raised 4 feet or more above the level of the main floor and shall include the area providing access to the seating area or serving only as a foyer.

BALCONY, EXTERIOR EXIT. See Section 3301 (b).

BASEMENT is any floor level below the first story in a building, except that a floor level in a building having only one floor level shall be classified as a basement unless such floor level qualifies as a first story as defined herein.

BOILER, HIGH-PRESSURE, is a boiler furnishing steam at pressures in excess of 15 pounds per square inch (psi) or hot water at temperatures in excess of 250°F., or at pressures in excess of 160 psi.

BOILER ROOM is any room containing a steam or hot-water boiler.

BUILDING is any structure used or intended for supporting or sheltering any use or occupancy.

BUILDING, EXISTING, is a building erected prior to the adoption of this code, or one for which a legal building permit has been issued.

BUILDING OFFICIAL is the officer or other designated authority charged with the administration and enforcement of this code, or the building official's duly authorized representative.

C

Sec. 404. CAST STONE is a precast building stone manufactured from portland cement concrete and used as a trim, veneer or facing on or in buildings or structures.

CENTRAL HEATING PLANT is environmental heating equipment which directly utilizes fuel to generate heat in a medium for distribution by means of ducts or pipes to areas other than the room or space in which the equipment is located.

C.F.R. is the Code of Federal Regulations, a regulation of the United States of America available from the Superintendent of Documents, United States Government Printing Office, Washington, D.C. 20402.

CHIEF OF THE FIRE DEPARTMENT is the head of the fire department or a regularly authorized deputy.

COMBUSTIBLE LIQUID. See the Fire Code.

CONGREGATE RESIDENCE is any building or portion thereof which contains facilities for living, sleeping and sanitation, as required by this code, and may include facilities for eating and cooking, for occupancy by other than a family. A congregate residence may be a shelter, convent, monastery, dormitory, fraternity or sorority house but does not include jails, hospitals, nursing homes, hotels or lodging houses.

CONDOMINIUM, RESIDENTIAL. See "apartment house."

CONTROL AREA is a space bounded by not less than a one-hour fire-resistive occupancy separation within which the exempted amounts of hazardous materials may be stored, dispensed, handled or used.

CORROSIVE is a chemical that causes visible destruction of, or irreversible alterations in, living tissue by chemical action at the site of contact. A chemical is considered to be corrosive if, when tested on the intact skin of albino rabbits by the method described in the United States Department of Transportation in Appendix A to C.F.R. 49 Part 173, it destroys or changes irreversibly the structure of the tissue at the site of contact following an exposure period of four hours. This term shall not refer to action on inanimate surfaces.

COURT is a space, open and unobstructed to the sky, located at or above grade level on a lot and bounded on three or more sides by walls of a building.

D

Sec. 405. DANGEROUS BUILDINGS CODE is the Uniform Code for the Abatement of Dangerous Buildings promulgated by the International Conference of Building Officials, as adopted by this jurisdiction.

DISPENSING is the pouring or transferring of any material from a container, tank or similar vessel, whereby vapors, dusts, fumes, mists or gases may be liberated to the atmosphere.

DISPERSAL AREA, SAFE. See Section 3322 (b).

DRAFT STOP is a material, device or construction installed to restrict the movement of air within open spaces of concealed areas of building components such as crawl spaces, floor-ceiling assemblies, roof-ceiling assemblies and attics.

DWELLING is any building or portion thereof which contains not more than two dwelling units.

DWELLING UNIT is any building or portion thereof which contains living facilities, including provisions for sleeping, eating, cooking and sanitation, as required by this code, for not more than one family, or a congregate residence for 10 or less persons.

E

Sec. 406. EFFICIENCY DWELLING UNIT is a dwelling unit containing only one habitable room.

ELECTRICAL CODE is the National Electrical Code promulgated by the National Fire Protection Association, as adopted by this jurisdiction.

ELEVATOR CODE is the safety code for elevators, dumbwaiters, escalators and moving walks as adopted by this jurisdiction (see Appendix Chapter 51).

EMERGENCY CONTROL STATION is an approved location on the premises of a Group H, Division 6 Occupancy where signals from emergency equipment are received and which is continually staffed by trained personnel.

EXISTING BUILDINGS. See "building, existing."

EXIT. See Section 3301 (b).

EXIT COURT. See Section 3301 (b).

EXIT PASSAGEWAY. See Section 3301 (b).

F

Sec. 407. FABRICATION AREA (fab area) is an area within a Group H, Division 6 Occupancy in which there are processes involving hazardous production materials and may include ancillary rooms or areas such as dressing rooms and offices that are directly related to the fab area processes.

FAMILY is an individual or two or more persons related by blood or marriage or a group of not more than five persons (excluding servants) who need not be related by blood or marriage living together in a dwelling unit.

FIRE ASSEMBLY. See Section 4306 (b).

FIRE CODE is the Uniform Fire Code promulgated jointly by the Western Fire Chiefs Association and the International Conference of Building Officials, as adopted by this jurisdiction.

FIRE RESISTANCE or **FIRE-RESISTIVE CONSTRUCTION** is construction to resist the spread of fire, details of which are specified in this code.

FIRE-RETARDANT-TREATED WOOD is any wood product impregnated with chemicals by a pressure process or other means during manufacture, and which, when tested in accordance with U.B.C. Standard No. 42-1 for a period of 30 minutes, shall have a flame spread of not over 25 and show no evidence of progressive combustion. In addition, the flame front shall not progress more than $10^1/_2$ feet beyond the center line of the burner at any time during the test. Materials which may be exposed to the weather shall pass the accelerated weathering test and be identified as Exterior type, in accordance with U.B.C. Standard No. 25-28. Where material is not directly exposed to rainfall but exposed to high humidity conditions, it shall be subjected to the hygroscopic test and identified as Interior Type A in accordance with U.B.C. Standard No. 25-28.

All materials shall bear identification showing the fire performance rating thereof. Such identifications shall be issued by an approved agency having a service for inspection of materials at the factory.

FLAMMABLE LIQUID. See the Fire Code.

FLOOR AREA is the area included within the surrounding exterior walls of a building or portion thereof, exclusive of vent shafts and courts. The floor area of a building, or portion thereof, not provided with surrounding exterior walls shall be the usable area under the horizontal projection of the roof or floor above.

FM is Factory Mutual Engineering and Research, 1151 Boston-Providence Turnpike, Norwood, Massachusetts 02062.

FOAM PLASTIC INSULATION is a plastic which is intentionally expanded by the use of a foaming agent to produce a reduced-density plastic containing voids consisting of hollow spheres or interconnected cells distributed throughout the plastic for thermal insulating or acoustical purposes and which has a density less than 20 pounds per cubic foot.

FOOTING is that portion of the foundation of a structure which spreads and transmits loads directly to the soil or the piles.

FRONT OF LOT is the front boundary line of a lot bordering on the street and, in the case of a corner lot, may be either frontage.

G

Sec. 408. GARAGE is a building or portion thereof in which a motor vehicle containing flammable or combustible liquids or gas in its tank is stored, repaired or kept.

GARAGE, PRIVATE, is a building or a portion of a building, not more than 1,000 square feet in area, in which only motor vehicles used by the tenants of the building or buildings on the premises are stored or kept. (See Chapter 11.)

GARAGE, PUBLIC, is any garage other than a private garage.

GRADE (Adjacent Ground Elevation) is the lowest point of elevation of the finished surface of the ground, paving or sidewalk within the area between the building and the property line or, when the property line is more than 5 feet from the building, between the building and a line 5 feet from the building.

GRADE (Lumber) is the classification of lumber in regard to strength and utility.

GUARDRAIL is a system of building components located near the open sides of elevated walking surfaces for the purpose of minimizing the possibility of an accidental fall from the walking surface to the lower level.

GUEST is any person hiring or occupying a room for living or sleeping purposes.

GUEST ROOM is any room or rooms used or intended to be used by a guest for sleeping purposes. Every 100 square feet of superficial floor area in a dormitory shall be considered to be a guest room.

H

Sec. 409. HABITABLE SPACE (ROOM) is space in a structure for living, sleeping, eating or cooking. Bathrooms, toilet compartments, closets, halls, storage or utility space, and similar areas, are not considered habitable space.

HANDLING is the deliberate transport of materials by any means to a point of storage or use.

HANDRAIL is a railing provided for grasping with the hand for support. See also Section 408, definition of "guardrail."

HAZARDOUS PRODUCTION MATERIAL (HPM) is a solid, liquid or gas that has a degree of hazard rating in health, flammability or reactivity of 3 or 4 and which is used directly in research, laboratory or production processes which have, as their end product, materials which are not hazardous.

HEALTH HAZARD is a classification of a chemical for which there is statistically significant evidence based on at least one study conducted in accordance with established scientific principles that acute or chronic health effects may occur in exposed persons. The term "health hazard" includes chemicals which are carcinogens, toxic or highly toxic agents, reproductive toxins, irritants, corrosives, sensitizers, hepatotoxins, nephrotoxins, neurotoxins, agents which act on the hematopoietic system, and agents which damage the lungs, skin, eyes or mucous membranes.

HEIGHT OF BUILDING is the vertical distance above a reference datum measured to the highest point of the coping of a flat roof or to the deck line of a mansard roof or to the average height of the highest gable of a pitched or hipped roof. The reference datum shall be selected by either of the following, whichever yields a greater height of building:

1. The elevation of the highest adjoining sidewalk or ground surface within a 5-foot horizontal distance of the exterior wall of the building when such sidewalk or ground surface is not more than 10 feet above lowest grade.

2. An elevation 10 feet higher than the lowest grade when the sidewalk or ground surface described in Item 1 above is more than 10 feet above lowest grade.

The height of a stepped or terraced building is the maximum height of any segment of the building.

HELIPORT is an area of land or water or a structural surface which is used, or intended for use, for the landing and take-off of helicopters, and any appurtenant areas which are used, or intended for use, for heliport buildings and other heliport facilities.

HELISTOP is the same as a heliport, except that no refueling, maintenance, repairs or storage of helicopters is permitted.

HIGHLY TOXIC MATERIAL is a material which produces a lethal dose or a lethal concentration which falls within any of the following categories:

1. A chemical that has a median lethal dose (LD_{50}) of 50 milligrams or less per kilogram of body weight when administered orally to albino rats weighing between 200 and 300 grams each.

2. A chemical that has a median lethal dose (LD_{50}) of 200 milligrams or less per kilogram of body weight when administered by continuous contact for 24 hours (or less if death occurs within 24 hours) with the bare skin of albino rabbits weighing between 2 and 3 kilograms each.

3. A chemical that has a median lethal concentration (LC_{50}) in air of 200 parts per million by volume or less of gas or vapor, or 2 milligrams per liter or less of mist, fume or dust, when administered by continuous inhalation for one hour (or

less if death occurs within one hour) to albino rats weighing between 200 and 300 grams each.

Mixtures of these materials with ordinary materials, such as water, may not warrant a classification of highly toxic. While this system is basically simple in application, any hazard evaluation which is required for the precise categorization of this type of material shall be performed by experienced, technically competent persons.

HORIZONTAL EXIT. See Section 3301 (b).

HOTEL is any building containing six or more guest rooms intended or designed to be used, or which are used, rented or hired out to be occupied, or which are occupied for sleeping purposes by guests.

HOT-WATER HEATING BOILER is a boiler having a volume exceeding 120 gallons, or a heat input exceeding 200,000 Btu/h, or an operating temperature exceeding 210°F. that provides hot water to be used externally to itself.

HPM STORAGE ROOM is a room used for the storage or dispensing of hazardous production materials (HPM) and which is classified as a Group H, Division 2, 3 or 7 Occupancy.

I

Sec. 410. IRRITANT is a chemical which is not corrosive but which causes a reversible inflammatory effect on living tissue by chemical action at the site of contact. A chemical is a skin irritant if, when tested on the intact skin of albino rabbits by the methods of 16 C.F.R. 1500.41 for four hours' exposure or by other appropriate techniques, it results in an empirical score of 5 or more. A chemical is an eye irritant if so determined under the procedure listed in 16 C.F.R. 1500.42 or other appropriate techniques.

J

Sec. 411. JURISDICTION, as used in this code, is any political subdivision which adopts this code for administrative regulations within its sphere of authority.

K

Sec. 412. No definitions.

L

Sec. 413. LINTEL is a structural member placed over an opening or a recess in a wall and supporting construction above.

LIQUID is any material which has a fluidity greater than that of 300 penetration asphalt when tested in accordance with the Uniform Fire Code Standards. When not otherwise identified, the term "liquid" is both flammable and combustible liquids.

LIQUID STORAGE ROOM is a room classified as a Group H, Division 3 Occupancy used only for the storage of flammable or combustible liquids in a closed condition. The quantities of flammable or combustible liquids in storage shall not exceed the limits set forth in the Fire Code.

LIQUID STORAGE WAREHOUSE is a Group H, Division 3 Occupancy used only for the storage of flammable or combustible liquids in an unopened condition. The quantities of flammable or combustible liquids stored are not limited.

LISTED and **LISTING** are terms referring to equipment and materials which are shown in a list published by an approved testing agency qualified and equipped for experimental testing and maintaining an adequate periodic inspection of current productions and which listing states that the material or equipment complies with accepted national standards which are approved, or standards which have been evaluated for conformity with approved standards.

LOADS. See Chapter 23.

LODGING HOUSE is any building or portion thereof containing not more than five guest rooms where rent is paid in money, goods, labor or otherwise.

LOW-PRESSURE HOT-WATER-HEATING BOILER is a boiler furnishing hot water at pressures not exceeding 160 psi and at temperatures not exceeding 250°F.

LOW-PRESSURE STEAM-HEATING BOILER is a boiler furnishing steam at pressures not exceeding 15 psi.

M

Sec. 414. MARQUEE is a permanent roofed structure attached to and supported by the building and projecting over public property. Marquees are regulated in Chapter 45.

MASONRY is that form of construction composed of stone, brick, concrete, gypsum, hollow-clay tile, concrete block or tile, glass block or other similar building units or materials or combination of these materials laid up unit by unit and set in mortar.

MASONRY, SOLID, is masonry of solid units built without hollow spaces.

MECHANICAL CODE is the Uniform Mechanical Code promulgated jointly by the International Conference of Building Officials and the International Association of Plumbing and Mechanical Officials, as adopted by this jurisdiction.

MEMBRANE PENETRATION FIRE STOP is a material, device or construction installed to resist, for a prescribed time period, the passage of flame, heat and hot gases through openings in a protective membrane in order to accommodate cables, cable trays, conduit, tubing, pipes or similar items.

MEZZANINE or **MEZZANINE FLOOR** is an intermediate floor placed within a room.

MOTEL shall mean hotel as defined in this code.

MOTOR VEHICLE FUEL-DISPENSING STATION is that portion of a building where flammable or combustible liquids or gases used as motor fuels are stored and dispensed from fixed equipment into the fuel tanks of motor vehicles.

N

Sec. 415. NONCOMBUSTIBLE as applied to building construction material means a material which, in the form in which it is used, is either one of the following:

(a) Material of which no part will ignite and burn when subjected to fire. Any material conforming to U.B.C. Standard No. 4-1 shall be considered noncombustible within the meaning of this section.

(b) Material having a structural base of noncombustible material as defined in Item (a) above, with a surfacing material not over $1/8$ inch thick which has a flame-spread rating of 50 or less.

"Noncombustible" does not apply to surface finish materials. Material required to be noncombustible for reduced clearances to flues, heating appliances or other sources of high temperature shall refer to material conforming to Item (a). No material shall be classed as noncombustible which is subject to increase in combustibility or flame-spread rating, beyond the limits herein established, through the effects of age, moisture or other atmospheric condition.

Flame-spread rating as used herein refers to rating obtained according to tests conducted as specified in U.B.C. Standard No. 42-1.

O

Sec. 416. OCCUPANCY is the purpose for which a building, or part thereof, is used or intended to be used.

ORIEL WINDOW is a window which projects from the main line of an enclosing wall of a building and is carried on brackets or corbels.

OWNER is any person, agent, firm or corporation having a legal or equitable interest in the property.

P

Sec. 417. PANIC HARDWARE. See Section 3301 (b).

PEDESTRIAN WALKWAY is a walkway used exclusively as a pedestrian trafficway.

PENETRATION FIRE STOP is a through-penetration fire stop or a membrane-penetration fire stop.

PERMIT is an official document or certificate issued by the building official authorizing performance of a specified activity.

PERSON is a natural person, heirs, executors, administrators or assigns, and also includes a firm, partnership or corporation, its or their successors or assigns, or the agent of any of the aforesaid.

PLASTIC MATERIALS, APPROVED, other than foam plastics regulated under Sections 1705 (e) and 1713, are those having a self-ignition temperature of 650°F. or greater and a smoke-density rating not greater than 450 when tested in accordance with U.B.C. Standard No. 42-1, in the way intended for use, or a smoke-density rating no greater than 75 when tested in the thickness intended for use by U.B.C. Standard No. 52-2. Approved plastics shall be classified and shall meet the requirements for either CC1 or CC2 plastic.

PLATFORM. See Chapter 39.

PLUMBING CODE is the Uniform Plumbing Code promulgated by the International Association of Plumbing and Mechanical Officials as adopted by this jurisdiction.

PROTECTIVE MEMBRANE is a surface material which forms the required outer layer or layers of a fire-resistive assembly containing concealed spaces.

PUBLIC WAY. See Section 3301 (b).

Q

Sec. 418. No definitions.

R

Sec. 419. REPAIR is the reconstruction or renewal of any part of an existing building for the purpose of its maintenance.

S

Sec. 420. SENSITIZER is a chemical that causes a substantial proportion of exposed people or animals to develop an allergic reaction in normal tissue after repeated exposure to the chemical.

SERVICE CORRIDOR is a fully enclosed passage used for transporting hazardous production materials and for purposes other than required exiting.

SHAFT is an interior space, enclosed by walls or construction, extending through one or more stories or basements which connects openings in successive floors, or floors and roof, to accommodate elevators, dumbwaiters, mechanical equipment or similar devices or to transmit light or ventilation air.

SHAFT ENCLOSURE is the walls or construction forming the boundaries of a shaft.

SHALL, as used in this code, is mandatory.

SMOKE DETECTOR is an approved device that senses visible or invisible particles of combustion.

STAGE. See Chapter 39.

STORY is that portion of a building included between the upper surface of any floor and the upper surface of the floor next above, except that the topmost story shall be that portion of a building included between the upper surface of the topmost floor and the ceiling or roof above. If the finished floor level directly above a usable or unused under-floor space is more than 6 feet above grade as defined herein for more than 50 percent of the total perimeter or is more than 12 feet above grade as defined herein at any point, such usable or unused under-floor space shall be considered as a story.

STORY, FIRST, is the lowest story in a building which qualifies as a story, as defined herein, except that a floor level in a building having only one floor level shall be classified as a first story, provided such floor level is not more than 4 feet below grade, as defined herein, for more than 50 percent of the total perimeter, or not more than 8 feet below grade, as defined herein, at any point.

STREET is any thoroughfare or public way not less than 16 feet in width which has been dedicated or deeded to the public for public use.

STRUCTURAL OBSERVATION means the visual observation of the structural system, including, but not limited to, the elements and connections at significant construction stages, and the completed structure for general conformance to the approved plans and specifications. Structural observation does not include or waive the responsibility for the inspections required by Sections 305 and 306.

STRUCTURE is that which is built or constructed, an edifice or building of any kind, or any piece of work artificially built up or composed of parts joined together in some definite manner.

SURGICAL AREA is the preoperating, operating, recovery and similar rooms within an outpatient health-care center.

T

Sec. 421. THROUGH-PENETRATION FIRE STOP is a material, device or construction installed to resist, for a prescribed time period, the passage of flame, heat and hot gases through openings which penetrate the entire fire-resistive assembly in order to accommodate cables, cable trays, conduit, tubing, pipes or similar items.

U

Sec. 422. U.B.C. STANDARDS is the Uniform Building Code Standards promulgated by the International Conference of Building Officials, as adopted by this jurisdiction. (See Chapter 60.)

UL is the Underwriters Laboratories Inc., 333 Pfingsten Road, Northbrook, Illinois 60062.

USE with reference to flammable or combustible liquids is the placing in action or service of flammable or combustible liquids whereby flammable vapors may be liberated to the atmosphere.

USE with reference to hazardous materials other than flammable or combustible liquids is the placing in action or making available for service by opening or connecting any container utilized for confinement of material whether a solid, liquid or gas.

USE, CLOSED SYSTEM, is use of a solid or liquid hazardous material in a closed vessel or system that remains closed during normal operations where vapors emitted by the product are not liberated outside of the vessel or system and the product is not exposed to the atmosphere during normal operations; and all uses of compressed gases. Examples of closed systems for solids and liquids include product conveyed through a piping system into a closed vessel, system or piece of equipment; and reaction process operations.

USE, OPEN SYSTEM, is use of a solid or liquid hazardous material in a vessel or system that is continuously open to the atmosphere during normal operations and where vapors are liberated, or the product is exposed to the atmosphere during normal operations. Examples of open systems for solids and liquids include dispensing from or into open breakers or containers, dip tank and plating tank operations.

V

Sec. 423. VALUE or VALUATION of a building shall be the estimated cost to replace the building and structure in kind, based on current replacement costs, as determined in Section 304 (b).

VENEER. See Section 3002.

W

Sec. 424. WALLS shall be defined as follows:

Bearing Wall is any wall meeting either of the following classifications:

(a) Any metal or wood stud wall which supports more than 100 pounds per lineal foot of superimposed load.

(b) Any masonry or concrete wall which supports more than 200 pounds per lineal foot superimposed load, or any such wall supporting its own weight for more than one story.

Exterior Wall is any wall or element of a wall, or any member or group of members, which defines the exterior boundaries or courts of a building and which has a slope of 60 degrees or greater with the horizontal plane.

Faced Wall is a wall in which the masonry facing and backing are so bonded as to exert a common action under load.

Nonbearing Wall is any wall that is not a bearing wall.

Parapet Wall is that part of any wall entirely above the roof line.

Retaining Wall is a wall designed to resist the lateral displacement of soil or other materials.

WATER HEATER is an appliance designed primarily to supply hot water and is equipped with automatic controls limiting water temperature to a maximum of 210°F.

WEATHER-EXPOSED SURFACES are all surfaces of walls, ceilings, floors, roofs, soffits and similar surfaces exposed to the weather, excepting the following:

(a) Ceilings and roof soffits enclosed by walls or by beams which extend a minimum of 12 inches below such ceiling or roof soffits.

(b) Walls or portions of walls within an unenclosed roof area, when located a horizontal distance from an exterior opening equal to twice the height of the opening.

(c) Ceiling and roof soffits beyond a horizontal distance of 10 feet from the outer edge of the ceiling or roof soffits.

X

Sec. 425. No definitions.

Y

Sec. 426. YARD is an open, unoccupied space, other than a court, unobstructed from the ground to the sky, except where specifically provided by this code, on the lot on which a building is situated.

Z

Sec. 427. No definitions.

10

Uniform Fire Code Definitions and Abbreviations

The following definitions are from Article 9 (Definitions and Abbreviations) of the 1991 Uniform Fire Code. *(Article 51 and 80 definitions and abbreviations follow.)*

"**BARRICADE** is a structure that consists of a combination of walls, floor and roof that is designed to withstand the rapid release of energy in an explosion. Barricades may be fully-confined, partially-vented or fully-vented."

"**CEILING LIMIT** is the maximum concentration of an airborne contaminant to which one may be exposed. The ceiling limits utilized are to be those published in 29 CFR 1910.1000."

"**CONTINUOUS GAS-DETECTION SYSTEM** is a gas-detection system where the analytical instrument is maintained in continuous operation and sampling is performed without interruption. Analysis may be performed on a cyclical basis at intervals not to exceed 30 minutes."

"**CONTROL AREA** is space within a building where the exempt amounts may be stored, dispensed, used or handled."

"**CYLINDER** is pressure vessel designed for pressures higher than 40 pounds per square inch, absolute, and having a circular cross section. It does not include a portable tank, multi-unit tank car tank, cargo tank or tank car."

"**EXCESS FLOW CONTROL** is a fail-safe system designed to shut off flow due to a rupture in pressurized piping systems."

"**EXCESS FLOW VALVE** is a valve inserted into a compressed gas cylinder, portable tank or stationary tank that is designed to positively shut off the flow of gas in the event that its predetermined flow is exceeded."

"**FABRICATION AREA (Fab Area)** is an area within a Group H, Division 6 Occupancy in which there are processes involving hazardous production materials and may include ancillary rooms or areas such as dressing rooms and offices that are directly related to the fab area processes."

"**HAZARDOUS PRODUCTION MATERIAL (HPM)** is a solid, liquid or gas that has a degree-of-hazard rating in health, flammability or reactivity of Class 3 or 4 as ranked by UFC Standard No. 79-3 and which is used directly in research, laboratory or production processes which have as their end product materials which are not hazardous."

"**HPM FLAMMABLE LIQUID** is an HPM liquid that is defined as being either flammable or combustible under the definitions listed in Article 9."

"**HPM STORAGE ROOM** is a room used for the storage or dispensing of HPM and which is classified as a Group H, Division 2, 3 or 7 Occupancy."

"**HEALTH HAZARD** is a classification of a chemical for which there is statistically significant evidence based on at least one study conducted in accordance with established scientific principles that acute or chronic health effects may occur in exposed persons. Health hazards include chemicals which are carcinogens, toxic or highly toxic materials, reproductive toxins, irritants, corrosives, sensitizers, hepatotoxins, nephrotoxins, neurotoxins, agents which act on the hematopoetic system, and agents which damage the lungs, skin, eyes or mucous membranes."

"**HIGHLY VOLATILE LIQUID** is a liquid with a boiling point of less than 68°F."

"**IDLH (Immediately Dangerous to Life and Health)** is a concentration of airborne contaminants, normally expressed in parts per million (ppm) or milligrams per cubic meter, which represents the maximum level from which one could escape within 30 minutes without any escape-impairing symptoms or irreversible health effects. This level is established by the National Institute of Occupational Safety and Health (NIOSH). If adequate data do not exist for precise establishment of IDLH data, an independent certified industrial hygienist, industrial toxicologist or appropriate regulatory agency shall make such determination."

"**INSIDE HPM STORAGE ROOM** is an HPM storage room totally enclosed within a building and having no exterior walls."

"**PERMISSIBLE EXPOSURE LIMIT (PEL)** is the maximum permitted eight-hour time-weighted average concentration of an airborne contaminant. The maximum permitted time-weighted average exposures to be utilized are those published in 29 CFR 1910.1000."

"**PHYSICAL HAZARD** is a classification of a chemical for which there is scientifically valid evidence that it is a combustible liquid, compressed gas, cryogenic, explosive, flammable gas, flammable liquid, flammable solid, organic peroxide, oxidizer, pyrophoric, unstable (reactive) or water-reactive material."

"**PORTABLE TANK** is any packaging over 60 U.S. gallons capacity and designed primarily to be loaded into or on or temporarily attached to a transport vehicle or ship and equipped with skids, mounting or accessories to facilitate handling of the tank by mechanical means. It does not include any cylinder having less than a 1,000-point water capacity, cargo tank, tank car tank or trailers carrying cylinders of over 1,000-pound water capacity."

"**REDUCED FLOW VALVE** is a valve equipped with a restricted flow orifice and inserted into a compressed gas cylinder, portable tank or stationary tank that is designed to reduce the maximum flow from the valve under full flow conditions. The maximum flow rate from the valve is determined with the valve allowed to flow to atmosphere with no other piping or fittings attached."

"**SEPARATE GAS STORAGE ROOM** is a separate enclosed area which is part of or attached to a building and is utilized for the storage or use of highly toxic compressed or liquefied gases."

"**STATIONARY TANK** is packaging designed primarily for stationary installations not intended for loading, unloading or attachment to a transport vehicle as part of its normal operation in the process of use. It does not include cylinders having less than 1,000-pound water capacity."

"**STORAGE FACILITY** is a building, portion of a building or exterior area used for the storage of hazardous materials in excess of exempt amounts specified in Division III."

"**USE (Material)** is the placing in action or making available for service by opening or connecting anything utilized for confinement of material whether a solid, liquid or gas."

The following definitions are from Article 51 (Semiconductor Fabrication Facilities Using Hazardous Production Materials) of the 1991 Uniform Fire Code.

"EMERGENCY ALARM SYSTEM is a system intended to provide the indication and warning of abnormal conditions and summon appropriate aid."

"EMERGENCY CONTROL STATION is an approved location on the premises of a Group H, Division 6 Occupancy where signals from emergency equipment are received and which is continually staffed by trained personnel."

"SERVICE CORRIDOR is a fully-enclosed passage used for transporting hazardous production materials and purposes other than required exiting."

"WORKSTATION is a defined space or an independent principal piece of equipment using hazardous production materials within a fabrication area where a specific function, a laboratory procedure or a research activity occurs. Approved cabinets serving the workstation are included as a part of the workstation. A workstation may contain ventilation equipment, fire protection devices, sensors for gas and other hazards, electrical devices and other processing and scientific equipment."

The definition below comes from Article 80 (Hazardous Materials) of the 1991 Uniform Fire Code.

"CONTAINER is any vessel of 60 U.S. gallons or less capacity used for transporting or storing hazardous materials."

Index